My Universe

A Theory of Yangton and Yington Pairs

Edward Wu

Preface

On March 3, 2017, I published the first edition of "My Universe – A Theory and Yangton and Yington Pairs" which covered my first 7 science papers related to Yangton and Yington Theory from 2015 to 2017. Since then, I have published another 18 papers in the following topics:

1. Vision of Light, Photon Inertia Transformation and Equation of Light Speed.
2. Hubble's Law and Universe Expansion derived from Redshift caused by Acceleration Doppler Effect.
3. Wu's Spacetime Field Equations based on Yangton and Yington Theory.
4. Standard Model and Quantum Field Theory versus Wu's Pairs and Yangton and Yington Theory.
5. Five Principles of the Universe and the correlations of Wu's Pairs and Force of Creation to String Theory and Unified Field Theory.
6. The true meanings of Mass, Time and Length and Principle of Correspondence.
7. Hubble's Law and Reverse Expansion Theories derived from Redshift caused by Wu's Spacetime Shrinkage.
8. Mass and Energy Conversion and Energy and Space Annihilation.
9. Einstein's Spacetime and Einstein's Field Equations versus Wu's Spacetime and Wu's Spacetime Field Equations.
10. Black Hole and Event Horizon interpreted by Photon Inertia Transformation and Wu's Spacetime Field Equations.
11. Einstein's Seven Mistakes resulted from his two wrong assumptions: (1) Light speed is always constant, and (2) Acceleration is the principle factor of the universe.

12. General Relativity Interpreted by Yangton and Yington Theory and Corresponding Identical Objects and Events in large gravitational field observed on Earth.
13. The correlations of Wu Constant and Wu's Spacetime Constant to Hubble Constant.
14. Derivations of Planck Constant, De Broglie Wave and Mass of Photon (Wu's Pair) Based on Yangton and Yington Theory.
15. Interpretation of Higgs Bosons as String Force Carriers and Higgs Field as the distribution of String Force and also that of Wu's Pairs.
16. Perihelion Precession of Mercury and Deflection of Light Interpreted by Yangton and Yington Theory.
17. A Summary of Wu's Spacetime Field Equation and Its Comparison to Einstein's Field Equation.
18. A summary and road map of Yangton and Yington Theory.

When a photon emitted from light source, it undergoes an Inertia Transformation where photon travels with two speeds, the Absolute Speed $3x10^8$ m/s (the speed of photon away from the light source) and the Inertia Speed (the speed of light source away from the observer). Light Speed is a vector summation of Absolute Light Speed and Inertia Light Speed.

According to the whirlpool model, where the momentum of a spinning particle is proportional to the mass and the spinning frequency of the particle, De Broglie Matter Wave, Planck constant and mass of Photon (Wu's Pair) can all be derived based on Yangton and Yington Theory.

Hubble's Law can be derived from the redshif caused by Acceleration Doppler Effect, which is used to explain the acceleration and expansion of the universe. However, the Dark Energy needed for the acceleration remains a mystery. Hubble's Law can also be derived from the redshift caused by Wu's Spacetime Shrinkage Theory such that Wu's Spacetime Reverse Expansion Theory can be used to interpret the expansion of the universe. Therefore, it is believed that the Spacetime on earth is actually shrinking instead of that the universe is expanding. In addition, Wu Constant K and Wu's Spactime γ can be calculated and expressed by Hubble Constant H_0.

In Wu's Spacetime Field Equation, the Amount of Normal Unit Acceleration (the curvature of the spacetime) is proportional to C^{-4} which is a function of Wu's Unit Length depending on the gravitational field and aging of the universe at the reference point. Because both G and C^{-4} appear on the matter and energy side (right hand side) of the equations, Einstein's Field Equations can be considered as a special case of Wu's Spacetime Field Equations observed on earth ($C_0 = 3x10^8$ m/s). In addition, Einstein's Field Equation focuses only on the correlation between the derivative of the curvature of space-time continuum and Amount of Normal Acceleration, where the space-time continuum is derived from a nonlinear geometry system to a Normal Spacetime System on earth. However, Wu's Spacetime Field Equation is derived upon the correlation between Amount of Normal Acceleration and gravitational field in Wu's Spacetime System on earth.

Standard Model is a group of subatomic particles derived by a mathematical model based on quantum field theory and Yang Mills Theory. In contrast, Wu's Pairs is a physical model proposed as the building blocks of all subatomic particles based on the Yangton and Yington Theory. Since all subatomic particles are made of Wu's Pairs, therefore Standard Model and quantum field theory should align with Wu's Pairs and Yangton and Yington Theory. As a result, Quantum Field Theory can be considered as a quantized field based on point particles and the distribution of particle radiation and contact interaction. Also, Quantum Gravity Theory can be interpreted as a quantized gravitational field based on the gravitons of string structure and the distribution of particle radiation and contact interaction. Furthermore, Unified Field Theory can be explained as a quantized field based on the Four Basic Forces between various strings structures that are made of Wu's Pairs with String Force induced from Force of Creation.

Based on a logical thinking, Five Principles of the Universe are proposed as the foundation of Wu's Pairs and Yangton and Yington Theory. These Five Principles include:

1. There was Nothing in the universe in the beginning.
2. From Nothing to Something it must be a reversible process.
3. The Something must be a pair of Antimatter particles with an inter-attractive force such that they can attract and destroy each other.
4. From Something to permanent matter there must be an external energy to cause a constant circulation motion between the two Antimatter particles so as to avoid them from recombination and destruction.
5. Eventually the whole universe will end and go back to Nothing.

During Big Bang Explosion, Space and Energy were first generated from None through Singularity. Then, Energy of Creation and Energy of Circulation combined together to form Wu's Pairs (Matter). Although Mass and Energy conversion can be commonly found in nuclear reaction, Einstein's $E=MC^2$, instead of mass and energy conversion, is actually an energy conversion between matter's structure energy generated from String Force and Four Basic Forces, and photon's kinetic energy.

Since Wu's Pairs are the building blocks of all matter, it is obvious that a fundamental measuring system can be established based on the Unit Mass – the mass of Wu's Pair (a pair of Yangton and Yington Circulating Particles), Unit Time – the period of the circulation of Wu's Pair, and Unit Length – the diameter of the circulation orbit of Wu's Pair.

When an object or event takes place or moves to a different location under an equilibrium condition, it maintains the same mass, structural feature and time sequence. In other words, it remains the same amounts of unit length and unit time, despite of the changes due to the gravitational field and aging of the universe. This object is called "Corresponding Identical Object", the event is called "Corresponding Identical Event" and the phenomenon is called "Principle of Correspondence". A corresponding identical object or event on a massive star (or black hole) has large length and time, but small velocity and acceleration because of the large gravitational field. When the same object and event is observed on earth, the Amounts of Unit Quantities measured by the Unit Quantities on earth are further enhanced due to the small gravitational field. This result agrees very well with Einstein's General Relativity.

According to Principle of Correspondence, both deflection of light and Perihelion Precession of Mercury are resulted from the decreasing speeds of photon and Mercury, while passing through a massive star (sun), due to the large Wu's Unit Length ($V \propto l_{yy}^{-1/2}$) caused by an extremely large gravitational force.

Wu's Spacetime and Wu's Spacetime Field Equations are based on Wu's Unit Time (t_{yy}) and Wu's Unit Length (l_{yy}) depending on the gravitational field and aging of the universe at the reference point. Einstein's Spacetime and Field Equations are only a special case of Wu's Spacetime and Wu's Spacetime Field Equations that are based on earth. Also, Black Hole and Event Horizon can be interpreted by Wu's Spacetime Field Equations and Photon Inertia Transformation.

Einstein claimed that light speed is constant; also time and length can change with acceleration. As a consequence, Einstein derived his theories including Special Relativity, General Relativity, Spacetime, Field Equations and Mass and Energy Conservation based on two wrong assumptions: (1) Light speed is always constant no matter the light source and observer, and (2) Acceleration is the principle factor of the universe.

In contrast, according to Yangton and Yington Theory, it is believed that (a) Light speed is not constant, instead it is the vector summation of Absolute Light Speed C and Inertia Light Speed, and (b) Gravitational field and aging of the universe are the principle factors of Wu's Spacetime instead of acceleration. As a consequence, the time and length of a corresponding identical object or event are a function of Wu's Unit Time (t_{yy}) and Wu's Unit Length (l_{yy}) depending on the gravitational field and aging of the universe no matter of the acceleration.

A flow chart "The Physics of Yangton and Yington Theory" is added as the road map for the systematic derivations of all the theories. It also serves as an overview to the correlations between the major physical phenomena and the Yangton and Yington Theory.

Finally, the mistakes of taking Higgs Boson as Graviton in some of my previous publications and editions of this book are corrected. To follow Standard Model, Higgs Boson is redefined as the force particle that generates and carries the string force between two Wu's Pairs so as to give "mass" to a string structure. Graviton, on the other hand, is redefined as the force particle that generates and carries gravitational force between two string structures.

The purpose of this edition is to establish a solid foundation to Yangton and Yington Theory in compliance with General Relativity and Quantum Field Theory. I wish more scientists will join me in the study of Yangton and Yington Theory and hopefully a proof of the existence of Wu's Pairs can be found in the near future.

Edward T. H. Wu
Los Angeles
February 25, 2020

To my beautiful daughters

Crystal and Diana

Thanks for their loves

and strong supports

To my grandson

Ronon

Thanks for the pleasures of

his cute little face and sweet smile

Introduction

"My Universe – A Theory of Yangton and Yington Pairs" is an extraordinary book of physical science with an unprecedented concept that bridges two major but conflicting subjects Quantum Field Theory and General Relativity in modern physics. It also explains and correlates almost everything in the universe, including space, time, energy and matter, as well as those substances from subatomic particles all the way to the boundary of the universe based on Wu's Pairs and a hypothetical Yangton and Yington Theory.

There are many ambiguities in the modern physics. For examples: Is photon a particle or a wave? Is there a unified field theory can explain all forces? Is string theory true? Is light speed constant? Does time change with speed and gravity? Is there a dark energy? What is dark matter? Is the universe expanding? And accelerating? What is spacetime? What is the black hole? Is there a wormhole? Can we do time travel? So on and so forth. To answer these questions, we need a breakthrough in particle physics. We need to know what the "God's Particles" the building blocks of the universe really are?

Wu's Pairs, a superfine Yangton and Yington circulating Antimatter particle pairs, are proposed as the building blocks of all matter. Force of Creation, the inter-attractive force between Yangton and Yington Pair, is the fundamental force of the string force and four basic forces complying with Unified Field Theory in the formation of all subatomic particles. String Theory is interpreted by the string structures of the subatomic particles built upon Wu's Pairs with string force. Photon is a free Wu's Pair escaped from a substance through a two stage separation and ejection process.

Gravitational force is induced from Force of Creation through the contact interaction between two gravitons. As a consequence, gravitational field and gravitational waves are caused by Graviton Radiation and Contact Interaction.

Einstein's Special Relativity and Velocity Time Dilation are challenged by the variable light speeds resulted from Photon Inertia Transformation and Vision of Light. Acceleration Doppler Effect is used to explain the Redshift phenomenon and to derive the Hubble's Law. In addition, Einstein's Law of Mass and Energy Conservation $E = MC^2$ is revised by energy transformation.

Space and Time are defined based on the circulation period and orbital diameter of Wu's Pairs. Principle of Correspondence is explained. Wu's Spacetime and Spacetime Shrinkage Theories are derived to explain the phenomena of the Cosmological Redshift and the Gravitational Redshift. Einstein's General Relativity and Gravitational Time Dilation are interpreted based on Wu's Spacetime Shrinkage Theory. Despite the Dark Energy and expansion of the universe, Wu's Spacetime Reverse Expansion Theory is proposed to interpret Hubble's Law and the expansion and acceleration of the universe. In addition, Wu's Spacetime Field Equation is derived in comparison to Einstein's Field Equations, which can be used to predict the existence of Black Hole and Event Horizon. Also, Corresponding Identical Objects and Events in large gravitational field observed on earth are studied. Finally all Einstein's mistakes including his postulates and theories are discussed.

Besides the science and physics, I would like to share the stories of my childhood curiosity and the experience of my horizontal and logical thinking with my readers, such that they can have a clear picture of the methodological background of this book. I would also like to encourage little boys and girls to ask questions without fear while improving their horizontal and logical thinking in finding answers. Furthermore, I hope this book will help young adults and grown-ups to get a better understanding of the universe by common senses and simple languages. Should this book be able to incubate some young genius minds to become future Einsteins, it will be my biggest rewards.

Biography

My name is Edward Tao Hung Wu. I was born in Taiwan in 1952. I graduated from Tsing Hua University in 1976 and received my Ph. D. from the Materials Science and Engineering Department of UCLA in 1983.

From 1983 to 1987 I was a research scientist at Rockwell Science Center. I went back to teach at UCLA as an associate professor from 1987 to 1990.

In 1991 I left UCLA and started two companies, Twin Bridge Software Corporation, dedicated to the development of window based multilingual input methods and PiezoTek Corporation devoted to the manufacture and marketing of an innovative piezoelectric ceramic transformer technology.

I met my wife Shen-Hui Ling in 1978, when we were graduate students of UCLA Engineering School and both lived in Hershey Hall, the UCLA graduate student's dormitory. In the summer of

1980, we were married in the Catholic Church across the street from Hershey Hall. Since 1982, we have lived in Westlake Village and Agoura Hills California, a suburb of Los Angeles. We have two daughters, Crystal and Diana. They both graduated from UCLA as well. In 2014 I took early retirement from my company PiezoTek. Since then I have been working as a volunteer for American Cancer Society, and devoting my spare time to my patented Minisolar concentrator technology and my Yangton and Yington Theory.

From 2015 to 2020, I have published a total of 25 science papers related to the Yangton and Yington Theory. They are unified and integrated to explain the formation of the universe and the correlation between space, time, energy and matter based on Wu's Pairs, a superfine Yangton and Yington circulating Antimatter particle pairs. Although only a hypothetical theory, it has successfully explained many major physical phenomena such as subatomic structures, unified field theory, string theory, light speed, De Broglie Wave, Planck constant, Einstein's relativities, gravitational waves, Hubble's Law, cosmological redshift, dark energy, spacetime, Black Holes, Event Horizon, Einstein's field equations and the expansion and acceleration of the universe.

On March 3, 2017, I published the first edition of my book "My Universe – A Theory of Yangton and Yington Pairs" to cover all of my previous studies on this subject. But the whole Yangton and Yington Theory were not completed until two year later. On November, 24, 2018, my 66th birthday, I published the second edition of "My Universe" including 15 papers in memory of my father Chi Sen Wu and my mother Kuang In Uang. However, the final version of "My Universe" with a total of 25 papers was not released until February, 2020 after I have completely recovered from my surgery. I hope this book will open up a new frontier to all scientists so that a fundamental understanding of the universe can be achieved.

To my dear wife

Shen-Hui Ling Wu

Millions of thanks for her

constant support

and

endless love

Table of Contents

A7. Redshift Caused by Acceleration Doppler Effect and Hubble's Law Based on Wu's Spacetime Shrinkage Theory

A8. Vision of Object, Vision of Light, Photon Inertia Transformation and Their Effects on Light Speed and Special Relativity

A9. Hubble's Law Interpreted by Acceleration Doppler Effect and Wu's Spacetime Reverse Expansion Theory

A10. Wu's Spacetime Field Equation Based on Yangton and Yington Theory

A11. Standard Model and Quantum Field Theory Versus Wu's Pairs and Yangton and Yington Theory

A12. Five Principles of the Universe and the Correlations of Wu's Pairs and Force of Creation to String Theory and Unified Field Theory

A13. Mass, Time, Length, Vision of Object and Principle of Correspondence Based on Yangton and Yington Theory

A14. A Summary of Yangton and Yington Theory and Their Interpretations on Subatomic Particles, Gravitation and Cosmology

A15. Hubble's Law Derived from Wu's Spacetime Shrinkage Theory and Wu's Spacetime Reverse Expansion Theory versus Universe Expansion Theory

A16. Einstein's $E = MC^2$ as Energy Conversion Instead of Mass and Energy Conservation and Energy and Space Annihilation Based on Yangton and Yington Theory

A17. Einstein's Spacetime and Einstein's Field Equations Versus Wu's Spacetime and Wu's Spacetime Field Equations

A18. Einstein's Seven Mistakes

A19. General Relativity versus Yangton and Yington Theory – Corresponding Identical Objects and Events in large Gravitational Field Observed on Earth

A20. The correlations of Wu Constant and Wu's Spacetime Constant to Hubble Constant

A21. Derivations of Planck Constant, De Broglie Matter Waves and Mass of Photon (Wu's Pair) from Yangton and Yington Theory

A22. Higgs Boson and Graviton Interpreted by String Force and String Structures Based on Wu's Pairs and Yangton and Yington Theory

A23. Event Horizon and Black Hole Interpreted by Photon Inertia Transformation and Yangton and Yington Theory

A24. A Summary of Wu's Spacetime Field Equation and Its Comparison to Einstein's Field Equation

A25. Perihelion Precession of Mercury and Deflection of Light Interpreted by Yangton and Yington Theory

A26. Refined Definitions in Real Numbers and Vectors and Proof of Field Theories

A27. Mathematical Methodologies in Physics and Their Applications in Derivation of Velocity and Acceleration Theories

A28. Nature Quantities and Measured Quantities

A29. Physical Meanings of Arithmetic Operations

A30. Constants and Constant Quantities in Physics

A31. Refraction and Deflection of Light

A32. Perihelion Precession

Prelude

In 2014, after early retirement, I began my research for answers to the questions that have bothered me throughout my life such as how did the universe begin, how was it formed, how will it end, and what are the relationships among space, time, energy and matter?

I wrote my first paper "Yangton and Yington - A Hypothetical Theory of Everything" in the fall of 2014 and submitted to PHYSICAL REVIEW LETTERS and several other well-known science journals. They were all rejected. One of the chief editors replied politely, "We regret to inform you that we have to turn down your paper because it is beyond our scope".

Finally, I submitted my paper to a small online publisher, Science Journal Publication. They accepted and published it in the SCIENCE JOURNAL OF PHYSICS on March 17, 2015 [Annex 1]. Since then, I have published a total of 25 papers listed in [Annex 1] to [Annex 25], all of them related to Yangton and Yington Theory. They can be found on the AMERICAN JOURNAL OF MODEN PHYSICS published by Science Publishing Group and IOSR JOURNAL OF APPLIED PHYSICS published by International Organization of Scientific Research, also on the following two websites: https://www.researchgate.net/profile/Edward_Wu14 and https://independent.academia.edu/EdwardWu10.

The name Yangton and Yington originates from yin-yang which has been a popular theory in China for thousands of years. The Chinese have always believed that the universe began from "Nothing" (0), then started with "Origin" (1), divided into "Yin and Yang" (−1 and +1), followed by "Four Phases", "Eight Sets", and finally ended with "All Matter" in the universe. However, the Chinese never explained yin-yang in quality and quantity. Yin-yang stayed only as a concept throughout the Chinese history.

My Yangton and Yington Theory is a theory that explains the

formations, structures and phenomena of the universe based on "Wu's Pair", a super fine "Yangton and Yington" circulating Antimatter particle pair with an inter-attractive "Force of Creation", that are generated from an empty space known as "Nothing". Although it is only a theory, the whole concept was developed upon the following "Five Principles of the Universe" with logical thinking:

1. There was Nothing in the universe in the beginning.

2. Nothing to Something must be a reversible process.

3. The Something must be a pair of Antimatter particles with an inter-attractive force such that they can attract and destroy each other.

4. From Something to permanent matter, there must be an external energy to cause a constant circulation motion between the two Antimatter particles so as to avoid them from recombination and destruction.

5. Eventually the whole universe will end and go back to Nothing.

Because the Big Bang Theory provides an ideal external energy to trigger the formation and circulation of Yangton and Yington Antimatter particle pairs, a hypothetical "My Universe" can be developed successfully based on the above Five Principles of the Universe – Foundation of Yangton and Yington Theory.

So far, I have accomplished the following tasks in my studies: I defined Wu's Pair, a Yangton and Yington circulating Antimatter circulating pair with an inter-attractive Force of Creation, as the building block of all matter. I proposed that photon is a "Free Wu's Pair" traveling in space. String Theory can be interpreted as that all subatomic particles have a string structure formed by a group of Yangton and Yington circulating Antimatter particle pairs stacking together with string force. Gravitational force is generated between two gravitons of string structures. I also figured the structures of electron, positron, neutron, proton, and recognized their relationships to electrical force, weak force and strong force. Furthermore, I explained unified field theory based on the four basic forces generated from the inter-attractive Force of Creation between Yangton and Yington particle pairs.

After I have developed a set of models to simulate various structures of subatomic particles and dark matter based on Wu's Pairs (Yangton and Yington circulating Antimatter particle pairs), I gave a mechanism to explain how a Wu's Pair can be separated from its parent substance to form a photon traveling at a constant speed (Absolute Light Speed C) from the light source. Then, appling the concepts of Vision of Light and Photon Inertia Transformation, I have developed an equation of light speed and proved that light speed can change with observers moving at different speeds and directions with respect to the light source. This opposes Einstein's Special Relativity.

Later, I proposed Particle Radiation and Contact Interaction Theory to explain Newton's Law of Universal Gravitation and gravitational waves. Also, I defined time and space by the period and diameter of Wu's Pairs; introduced the Principle of Correspondence to correlate the properties of the corresponding identical objects and events; and interpreted their differences observed on earth.

Then, I applied Acceleration Doppler Effect to explain Doppler Redshift, Hubble's Law and the Universe Expansion. I also interpreted Cosmological Redshift, Gravitational Redshift and Hubble's Law based on Spacetime Theory and Spacetime Shrinkage Theory without Dark Energy. I derived Spacetime Reverse Expansion Theory to explain the expansion and acceleration of the universe. In addition, I explained Einstein's Law of Mass and Energy Conservation $E=MC^2$ based on Yangton and Yington Theory; proposed Wu's Spacetime Field Equation as a general form of Einstein's Field Equations and proved the existence of Black Hole and Event Horizon. Lastly, I discussed all Einstein's mistakes in his postulates and theories in details.

Furthermore, based on the whirlpool model, I have derived Planck Constant, De Broglie Wave and Mass of Photon (Wu's Pair). I have also successfully correlated Wu Constant, Wu's Spacetime Constant to Hubble Constant based on Yangton and Yington Theory.

As a result, Wu's Pairs and the Yangton and Yington Theory can be used successfully in derivation and explanation of many major physical phenomena and theories. A flow chart of their correlations is shown in Fig. G.

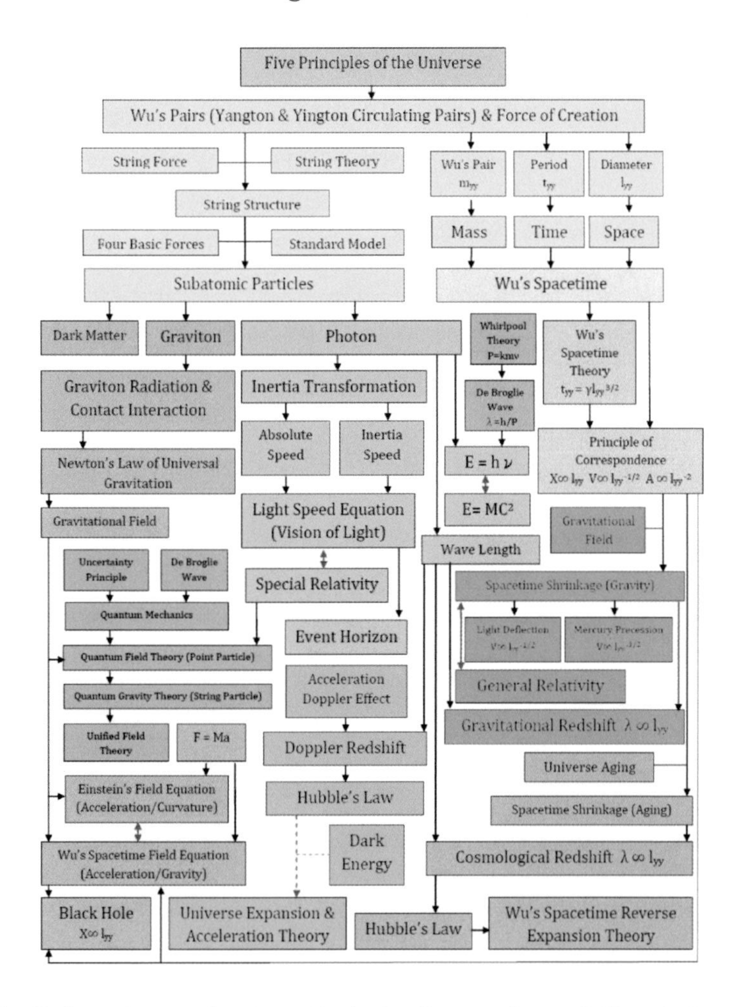

Fig. G A road map of systematic derivations and correlations between major physical phenomena and Yangton and YingtonTheory.

Although there is no proof of the existence of Wu's Pairs and Yangton and Yington Theory, one may consider Yangton and Yington Theory as a Binary Universe. Just like the relationship between Binary System and Decimal System in the mathematics, many theories and principles developed in the Binary Universe can be used effectively in reflection of the real universe.

Finally, in spite of science, I would like to share with my readers the stories of my childhood curiosity and the experience of my horizontal and logical thinking, such that they can have a clear picture of the background of this book. I hope to encourage little boys and girls to ask questions without fear while improving their horizontal and logical thinking in finding answers. I hope this book will also help many young adults and grown-ups to get a better understanding of our universe by some common senses and simple languages. Should this book be able to incubate some young genius minds to become our future Einsteins, it will be my biggest reward.

After all, I would like to give my deep gratitude to my dear friend Dan Witzling and my daughter's godparents Della and Gino Spinelli for their great efforts and kind assistances in reviewing and editing this book. I would also like to give my sincere appreciation to Dr. Louis Newman, for his comments and advices. Furthermore, I would like to thank my daughters Crystal and Diana for their positive suggestions and encouragement during the entire period of writing this book. Last but not least, I would like to give my deepest thanks to my wife Ling Wu (Shen Hui). Without her constant support and endless love, I would never be able to write and complete this book.

Chapter One

Questions without Answers

Why A + B = B + A?

Chicken or Egg, Which One Came First?

1.1. A Boy with a Big Curiosity

I am a Chinese immigrant in United States. My name is Edward Tao Hung Wu. People call me Edward and my Chinese friends call me Tao Hung. My parents used to call me by my nick name Hung, but I prefer to call myself Tao, which means the great way.

I was born in Taiwan in 1952. When I was young, I was the kind of boy who was always curious of everything. I often asked questions such as "Why the earth is round?", "Why dog and cat have tails but we don't?", "Chicken or egg, which one came first?" and even "Why one plus one equals to two?"

My father was an officer in the Taiwan military. He was a college drop-out from Nanjing University back in China during World War II. When I was in elementary school, he retired from the military and started teaching Chinese literature in a local girl's high school. He taught me how to play the Chinese game "Go", but he couldn't help me with my questions.

My mother was a house wife for her whole life and she never entered any school. However, she taught herself how to read and write. Of course, she too couldn't answer my questions.

I have two brothers and one sister. I am the youngest child in my family. They are all much older than me. My oldest brother, Tao An, was 21 years old, second brother, Tao Chung, was 18 years old, and my sister, Rho Lin, was 15 years old when I was born.

They always teased me and said that I was adopted from the homeless lady who lived in the bomb shelter down the hill close to the end of our village. It made me very mad. When I was in elementary school, they were all grown up and left home. Nobody in my family could teach me and answer my questions.

When I entered school, I started asking my teachers the same questions. Again, I couldn't get any answers. One day when I was in 9th grade, I could not understand what an infinitesimal ΔX really meant. I hung myself upside down on a lift bar in the school gym, thinking that it could help me to understand if I saw the world totally opposite. Of course, it didn't work and only showed how desperate and hopeless I was as a young scientist trying to find answers to my questions.

In 1968, after graduating from Miaoli Middle School, I entered Taipei Jianguo High School, the No. 1 boy's high school in Taiwan. I left my parents and went to live with my big brother, Tao An, and his wife in Taipei. I was a 15 year old country boy and quickly got lost in the neon signs of the metropolitan city. In the first year I was doing okay and could still maintain my grade in the top 30% of my class. However in my second year, my grade quickly dropped to the bottom 10% of the class. I had too many unanswered questions in my physics and mathematics classes. I couldn't focus on my studies. I was very upset and felt hopeless. In my senior year I told myself I needed to stay focused, forget about all the questions, and just take them for granted. I studied 16 hours a day nonstop for the whole year until the day of the national college entrance exam.

My hard work paid off. I was ranked in the top 1,000 students in the national college entrance exam. In September 1971, I entered the mathematics department of Tsing Hua University, which is one of the top three universities in Taiwan. I was so happy and promised myself that finally all my questions would get answered.

I remembered, the first day of my Number Theory class. I raised my hand and asked my young professor this question:

"Professor Tsai, how can we prove that A + B = B + A in Real Numbers?"

Professor Tsai was a fresh Ph.D. graduate from Perdue University. She thought a little while and answered,

"The Commutativity is a basic property of Field Theory and Field Theory is the principle of Real Numbers. It is a law and there is no proof. If you really want to prove it, you may try to plug in all real numbers into the formula and you can always find A + B equals to B + A, such as 3 + 5 = 8 and 5 + 3 = 8, also 1 + 6 = 7 and 6 + 1= 7, so on and so forth".

I said, "Professor, but there are infinite numbers and one can never prove them all".

She answered, "Yes, so it is better to just take it for granted".

I was very upset with her answer, even more so in the physics class when I asked Professor Lee, "How can we prove Newton's Law of Universal Gravitation $F = k (M_1M_2)/r^2$?"

Professor Lee was a very famous physics professor in the school. He had a very strong opinion about student's scores and he never gave his students a score of more than 60 points. Here is what he always said to his students, "There are so many geniuses in physics, if you can't get 60 points from my class, don't waste your time, and better change your major right away".

He stared at me and said, "It is a law. We can only prove it by experiments".

I asked, "Do you mean that we have to measure the gravity between all objects? It is impossible."

He answered, "Yes, or otherwise just take it for granted".

Again, there was no answer. The only thing I could do was "Take it for granted". It drove me crazy. One time, I seriously considered dropping out of college. Fortunately, I soon found out that there was a new department in the school Materials Science and Engineering that focused on the studies of properties and characterizations of materials. In this field, there would be minimal mathematics and physics. In the fall of 1973, I successfully transferred to the new department. It was a big relief of my nightmares dealing in physics and mathematics.

1.2. A Rising Materials Scientist

In 1976 I graduated from Tsing Hua University. After spending two years in mandatory military service in Taiwan, I came to United States and entered the University of California, Los Angeles (UCLA) in September of 1978, pursuing my Master and Ph. D. degrees in materials science and engineering.

After 5 years of hard work, I received my Ph. D. in the fall of 1983. I started my career as a research scientist at Rockwell Science Center in Thousand Oaks, California, where I grew single crystals for optical computer applications. In 1987, recommended by my Ph. D. thesis advisor Professor John D. Mackenzie, I went back to UCLA, working as an associate professor. I taught undergraduate and graduate courses while conducting research on sol gel technology, a pioneer work of nanometer technology.

In 1991 I left UCLA and started two companies, Twin Bridge Software Corporation dedicated to the development of Windows based multilingual input methods, and PiezoTek Corporation devoted to the manufacture and marketing of an innovative Piezoelectric Ceramic Transformer technology.

In 1992 Twin Bridge Software Corporation introduced the first Chinese, Korean and Japanese multilingual systems for Microsoft's Windows operation system. In 1996 TwinBridge again took the lead in launching Chinese OCR, Hand Writing and Voice Recognition software to the international markets.

In 1995 PiezoTek Corporation introduced the first Step-Up Piezoelectric Ceramic Transformer backlight power supply to the global LCD industry. Additionally, in 2005, PiezoTek developed the Step-Down Piezoelectric Ceramic Transformer charger and adaptor for commercial electronic devices. PiezoTek further announced the first worldwide patented Piezoelectric Ceramic Coupler for solid state relay applications in 2007.

In 2014, I retired from PiezoTek. Since then I have been working as a volunteer for the American Cancer Society, while spending my spare time devoted to my patented Minisolar Concentrator technology and my Yangton and Yington Theory.

I have always been passionate about been a materials scientist. I have been amazed by the beauty of crystals since the first time I learned about the crystalline structure. All gem stones, laser rods and semiconductors, not only have a perfect structure inside, but also a brilliant appearance outside. Most important, their fascinating properties made a revolution in human modern civilization. I love crystals so much that I even named my first daughter "Crystal".

A material scientist is different from a physicist or a mathematician. Both the physicist and the mathematician use mathematical formulas and tools to study and describe the universe. Like a blind man, he try to figure out an elephant's look by touching the elephant's nose, ears, teeth and trunks, and examining the elephant's responses by throwing stones at it. A material scientist, on the other hand, uses a physical model to study and simulate the universe.

Physicists have spent more than one century trying to understand the structure of atoms by bombarding them with particles and radiations until they finally found out that the atom is like a hollow balloon with electrons circulating on the surface and a tiny nucleus made of protons and neutrons at the center.

Similarly, in recent years, physicists studied the structure of subatomic particles by the Large Hadron Collider (LHC). Photon and more than a dozen quarks, leptons and Gauge Bosons were revealed during the collision. Since these particles are extremely small, finding the structures and what they are made of remains a big challenge.

However, with logical thinking, a simple particle model can be established. Assuming there are "X Particles" which are the basic building blocks of all matter, and then obviously a free X Particle (like photon) and a group of clusters of X Particles (like quarks), they could be released and detected during the particle collision. In other words, not only is the "X Particles" the building blocks, but the whole universe could be interpreted and structured based on the "X Particle" theory, which I named later in this book as "Wu's Pairs" and "Yangton and Yington Theory".

Before I explain the details of "My Universe" which is made of "Wu's Pairs" based on "Yangton and Yington Theory", I would like to give my readers insight into my thinking and the methodologies used in my studies.

1.3. Why the Earth Is Round?

This is the question I had when I was 7 years old. The answer is very simple. Since the sun, moon and stars are all round, therefore, the earth must also be round. That answer came from my logical thinking when I was at age 12.

Our ancestors might say that the earth is flat and endless. If I was born in ancient times I couldn't argue with them if the earth is flat or not (no proof at that time), but I certainly wouldn't agree that it is endless. Because there is a sunrise and sunset each day, which means that the Sun must go around the earth once every day. Therefore, the earth must be a finite object and can't be endless. As to whether or not the earth is round, I could guess that the earth is round just like all the stars in the sky. It is logic thinking, similar to that I could always say Mr. Smith has two hands even I never met him.

1.4. Chicken or Egg, Which One Came First?

The answer is the egg. Why? An egg contains a large cell that has only one set of DNA. It is different from that of a chicken, which contains many billions of cells each carrying the same DNA. From an egg to a chicken, through duplication, all the cells of the chicken must have exactly the same DNA as that of the parent cell in the egg. This is a much easier production process than the other way around. However, where the first chicken egg came from remains a question.

I believe that the DNA of the first chicken egg was given by two birds from the rearrangement of their DNA structures through breeding. Therefore, the egg came first makes more sense.

1.5. Why A + B = B + A?

The answer is in the physics, "The Law of Conservation of Mass". Why? Number A is the amount of a group of elements that is counted by a specific numerical sequence. For example three oranges means that the amount of a group of oranges is three, which is counted by an integer sequence: 1, 2, 3, 4, 5... etc. Summation "+" is an operation in which two groups of elements add together to form a new group and the amount of the new group is the corresponding number of the summation.

Both A + B and B + A have the same amount, because the total amount of elements in two groups maintains unchanged no matter what sequence it is counted. As a result, A + B = B + A is ensured by physics "The Law of Conservation of Mass" [Annex 26].

1.6. Horizontal Thinking

To solve a problem, we often find answers outside of the box instead of taking a straight approach. This is known as "Horizontal Thinking" or "Parallel Thinking". If one can only take a straight approach such as that a chicken delivers an egg and an egg grows up to become a chicken, then he will get himself into a controversy and can never find an answer. The answer egg came first is a result of horizontal thinking.

Another example is in the proof of "A + B = B + A". If one can only take the straight approach by plugging all different numbers into the formula, he will never be able to prove it. Using the physical principle "The Law of Conservation of Mass" to solve the problem instead of mathematical calculation, the answer again comes from horizontal thinking.

When I first worked on the Yangton and Yington Theory, I got myself into a dilemma between the reversible processes "Nothing can become Something" and "Something can go back to Nothing". Which one comes first? And how they work? I solved the problem by applying a horizontal thinking. I created a pair of Antimatter particles, the Yangton and Yington pair, with an inter-attractive force which is Something that can be generated from Nothing and also they can recombine and destroy each other to go back to Nothing.

1.7. Vertical Thinking

Besides horizontal thinking, which can initiate a creation, "Vertical Thinking", also known as "Logical Thinking", is the most important skill in finding a result, getting an answer, and solving a problem. Logical thinking requires a thorough analysis of a big data base. For example, "All stars and planets are round, therefore earth must be round" is a result of logical thinking. All stars and planets represent a big data base and they all have a round spherical shape is a result of analysis. Since earth is a planet, therefore earth must be round is a result of vertical thinking. Similarly, "All men have two hands, therefore Mr. Smith must have two hands" is also a result of logical thinking.

Everything must start from a simple form before it turns into a complex structure. For example, life starts from a unicellular organism long before the evolution to form a complex human being. Also, a chicken must start from a single cell egg to grow into a full size rooster. Because of this reasoning, the universe cannot start from beginning a complex structure such as a dozen quarks and leptons. That is why I developed Wu's Pairs, a Yangton and Yington circulating Antimatter particle pairs, as the building blocks to structure the whole universe. In general, a logical thinking can be used as a fundamental method in study of the universe [Annex 27].

1.8. A Little Imagination

Just like most kids, when I was young I had a lot of dreams and imaginings. I often dreamed of flying a F104 jet fighter and engaged in a dogfight with enemy's airplanes. I was so excited firing the machine gun and following the trace of the bullets until my enemy's airplanes were destroyed. Many years later, it was this trace of bullets that gave me the idea of Vision of Light. I used this concept to prove that light speed is not always constant. It changes with the observers at different speeds and different directions. These facts conflict with Einstein's Special Relativity and Velocity Time Dilation theory, in which light speed is postulated always a constant.

I have also used my imagination in building a model in explanation of the propagation of Gravitational Force. First, based on logical thinking, a messenger such as Graviton is needed, which has a string structure that is capable of generating and delivering gravitational force. Secondly, use an imagination such as that in the movie "Star Wars", where space fighters (Gravitons) are sent out from earth (parent object) to bombard the Death Star (target object). The total number of space fighters sent out from earth is proportional to the size of earth. The total number of laser guns on the Death Star is also proportional to the size of the Death Star. However, because of the radiation, the number of space fighters that could engage crossfire with the laser guns on the Death Star is reduced inversely proportional to the square of the distance between earth and the Death Star. Therefore, the total number of the crossfire (gravitational force) is proportional to the size of earth and the Death Star, which is also inversely proportional to the square of the distance between earth and the Death Star.

Another example is in the expansion of the universe. When I first heard that the universe is not only expanding but also accelerating, I was so confused. Because it was totally against my logical thinking, how can the universe expand at a speed even faster than light speed and how can the expansion accelerate without external energy. Could it be true, as many scientists suggested that there is a free Dark Energy in the universe that can be used to pump up the acceleration?

Again, I used my imagination to find the answers. I remembered that some years ago, I watched the movie "Honey, I shrunk the kids", in which all the subjects in the normal world such as people, houses, cars, furniture, dogs, cats and even ants became huge monsters in the eyes of the shrunken kids. Therefore, I believe that a similar mechanism happened in the universe. Instead of that universe expanding, our earth is actually shrinking. This imagination helped me a great deal in understanding and developing my Spacetime Shrinkage Theory to explain the expansion and acceleration of the universe.

Chapter Two

A Theory of Yangton & Yington Pairs

How the Universe Started?

What Is the Building Block of Matter?

2.1. Could Quarks Be the Basic Building Blocks of the Universe?

Everything should start from simple form rather than a complex one. Taking more than a dozen subatomic particles such as quarks and leptons as the building blocks of the universe is unbelievable. It is totally against our experience and common sense. Therefore, it is my belief, something simple such as photons should be the basic building blocks of all the matter in the universe.

Although taking photons as the building blocks of the universe sounds crazy, ask yourself why we can find photons everywhere in the universe such as that in the thermal radiation, nuclear reaction, electron oscillation, particle collision, and even in the early stages of the Big Bang explosion. If indeed that the photons are the building blocks of all matter, then what the photon structure is and how they glue themselves together to form subatomic particles become a big challenge to all scientists.

2.2. Yangton and Yington Theory – A Theory of Everything

Yangton and Yington Theory is a hypothetical theory based on a pair of super fine Antimatter particles named "Yangton and Yington" with an inter-attractive force named "Force of Creation" forming a permanent circulating particle pair named "Wu's Pair" that is proposed as the fundamental building blocks of the universe. The theory explains the formation of all the substances in the universe and the correlations between space, time, energy and matter.

2.3. Five Principles of the Universe – A Theory of Creation

According to Yangton and Yington Theory, it is proposed that a Wu's Pair containing two super fine Antimatter particles Yangton and Yington circulating in an orbit with an inter-attractive Force of Creation is generated from Nothing. Although it is just a hypothetical theory, the whole concept was developed based on the following Five Principles of the Universe [59] with a series of systematic logical thinking:

1. There was Nothing in the universe in the beginning.
2. From Nothing to Something it must be a reversible process.
3. The Something must be a pair of Antimatter particles with an inter-attractive force such that they can attract and destroy each other.
4. From Something to permanent matter there must be an external energy to cause a constant circulation motion between the two Antimatter particles so as to avoid them from recombination and destruction.
5. Eventually the whole universe will end and go back to Nothing.

To begin the introduction of the Yangton and Yington Theory, let's first start from the above five principles:

The 1st principle:

"There was Nothing in the universe in the beginning." This is the result of logical thinking. Otherwise, if the universe started from Something then one will always ask where that Something came from.

The 2nd principle:

"From Nothing to Something it must be a reversible process." This is also a result of logical thinking. Common sense tells us that everything that has a beginning must have an end. The question is how it ends? And how long it takes to end? Would it make more sense just to reverse the initial process from Something back to Nothing? Simply because that Nothing already existed and also the reverse could happen instantly at an equilibrium condition. Therefore, I believe that from Nothing to Something must be a reversible process.

The 3rd principle:

"The Something must be a pair of Antimatter particles with an inter-attractive force such that they can attract and destroy each other." As a result of logical thinking, the only possibility that Something can go back to Nothing is that the Something must have a built-in self destruction mechanism such as a pair of Antimatter particles Yangton and Yington Pair, with an inter-attractive Force of Creation for the enforcement of self destruction.

The 4th Principle:

"From Something to permanent matter there must be an external energy to cause a circulation motion between the two Antimatter particles so as to avoid recombination and destruction." A circulation motion between two particles can be produced by two opposite motions against a vertical force. Since Force of Creation is the vertical force between the two Antimatter particles, and the external force could be provided by Big Bang explosion, this principle is very well supported by the Big Bang Theory.

The 5th Principle:

"Eventually the whole universe will end and go back to Nothing." With logical thinking the universe can only be ended with Nothing. Otherwise, it will become a never ending story. Only going back to Nothing can stop this paradox.

2.4. Yangton and Yington – The Basic Particles

According to the 1st Principle and 2nd Principle of the Five Principles of the Universe, it is proposed that Yangton and Yington [1], a pair of super fine Antimatter particles can only be produced together with an inter-attractive Force of Creation simultaneously from an empty space called "Nothing", which is known as the 3rd Principle. This Yangton and Yington Pair with Force of Creation [1] called "Something" can recombine and destroy each other so that Something can go back to Nothing, which obeys 2nd Principle. Both Yangton and Yington are the fundamental particles of the universe. They can be used to form a Something (Fig. 1) as that in 3rd Principle and also a Wu's Pair (Fig. 2) as that in 4th Principle [1]. Something is only a temporary particle, but Wu's Pair is a permanent particle which is the building block of all matter such as photons, quarks, electrons, positrons, neutrons, protons and Dark Matters, etc [2].

Instead of solid particles, Yangton and Yington can also be considered as two tiny Energy Whirlpools (Energy Particles) with opposite spin up (Yangton) and spin down (Yington) directions generated by the energy released from the Big Bang explosion.

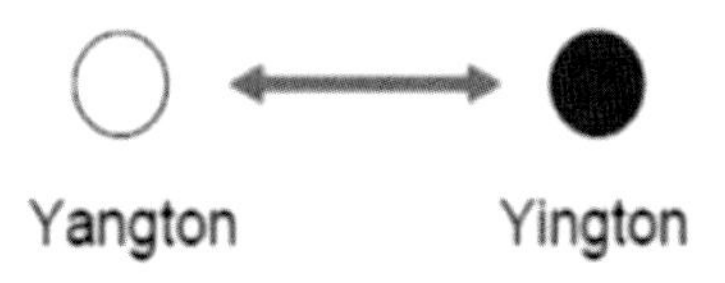

Fig. 1 Something - a Yangton and Yington pair with Force of Creation.

2.5. Force of Creation – The Fundamental Force

According to 2nd Principle and 3rd Principle, Yangton and Yington must coexist with an inter-attraction force named "Force of Creation" (Fig. 1), such that recombination and destruction can be enforced and Something will go back to Nothing. Therefore, the reaction of this reversible process can be represented by the following formulas:

Nothing → Yangton Θ Yington $\Delta E = E_{Creation}$

$E_{Creation}$ ↔ Yangton Θ Yington

Where "Θ" represents Force of Creation, "Yangton Θ Yington" represents Something and $E_{Creation}$ is Energy of Creation.

The inter-attractive Force of Creation between Yangton and Yington is the fundamental force in the universe, which can be used to generate String Force for the formation of elementary subatomic particles such as quarks, leptons, gluons and bosons; as well as the Four Basic Forces including gravitational force, electromagnetic force, weak force and strong force for the formation of composite subatomic particles such as proton, neutron and nucleus.

2.6. Big Bang – How the Universe Started?

About 13.8 billion years ago, there was nothing – no space, time, energy or matter, which is known as "None". Then a Big Bang [3] exploded. Immediately, space was created and energy was released from a single point known as "Singularity". Energy released from Big Bang explosion was used to generate Yangton and Yington Pairs with inter-attractive Force of Creation and subsequently drove them into a circulating motion. This circulating motion could prevent the recombination and destruction of the Yangton and Yington Pairs such that Something couldn't go back to Nothing and thus a permanent Wu's Pairs (Fig. 2) [1] could be formed.

2.7. Circulation – How Matter Become Permanent?

The energy released from the Big Bang explosion could drive Yangton and Yington particles into a circulating motion [1]. This circulating motion not only prevents the attraction and destruction between Yangton and Yington particles, but it also makes them alive and in operation.

Circulation can also be found commonly in our cosmos such as that electron circulating the nucleus, moons circulating planets, planets circulating stars, stars circulating the galaxies, etc. Therefore, circulation is the key to making a matter permanent. This is, again, a result of logical thinking.

2.8. Wu's Pair – The Building Block of the Universe

According to the 4th Principle, with the external energy generated from Big Bang explosion, a Yangton and Yington circulating pair with an inter-attractive Force of Creation named "Wu's Pair" (Fig. 2) can be formed so that Something can become a permanent matter. These Wu's Pairs are the fundamental building blocks (God's Particles) of all matter such as photons, quarks, electrons, positrons, neutrons, protons, etc.

From Something to a permanent Wu's Pair, the reaction process can be represented by the following formulas:

$$\text{Yangton } \Theta \text{ Yington} \rightarrow \text{Yangton } \Phi \text{ Yington} \quad \Delta E = E_{Circulation}$$

$$E_{Creation} + E_{Circulation} \leftrightarrow \text{Yangton } \Phi \text{ Yington}$$

Where "Yangton Θ Yington" represents Something – a temporary Yangton and Yington pair. "Yangton Φ Yington" represents Wu's Pair – a permanent Yangton and Yington circulating pair. $E_{Creation}$ is Energy of Creation which is used to generate Force of Creation. $E_{Circulation}$ is the circulation energy which includes both potential and kinetic energies of the circulation. The summation of $E_{Creation}$ and $E_{Circulation}$ is called "Wu's Pair Formation Energy" which can be generated either from Big Bang explosion or nuclear reaction.

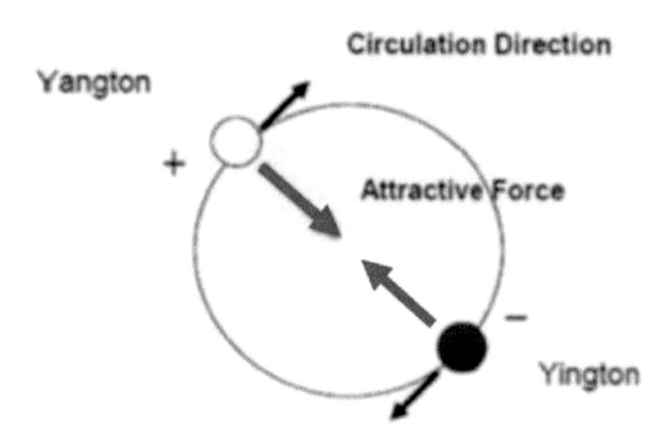

Fig. 2 Wu's Pair - a Yangton and Yington circulating pair.

2.9. Photon – A Free Wu's Pair

When Wu's Pair is released from a substance, it becomes a free particle known as "Photon" traveling in space at a constant speed 3×10^8 m/s observed at the light source. The reaction process can be represented by the following formula:

Yangton Φ Yington → Photon $\Delta E = h\nu$

"Yangton Φ Yington" is Wu's Pair and $h\nu$ is photon's kinetic energy.

2.10. Dancing Couple

Yangton and Yington are like to a man and a woman. God created them from Nothing. Man and Woman are born to love each other just like "Force of Creation" attracts Yangton and Yington together. When man and woman start dancing the waltz it is like Yangton and Yington beginning the circulation. As the attractive force is balanced by the centrifugal force while the whole world is in peace and harmony, everything turns to permanent.

I like to use the above example to explain my Yangton and Yington Theory. My audience always loved it and agreed with my theory no matter who they were and what religion they believed.

Chapter Three

Subatomic Particle Structures and Four Basic Forces

What Is the String Theory?

What Are the Structures of Subatomic Particles?

What Is the Gravitational Force?

What Is the Unified Field Theory?

3.1. String Theory, String Force and String Structure

When I first heard about String Theory, I wondered how it could be possible that all matter has a linear structure, like spaghetti. From my materials science background I know that only polymers and glass have a linear structure, but not those of single crystals, metals and inorganic compounds, which all have crystalline structures such as cubic, tetragonal and hexagonal, etc., in other words, a point structure. However, in order to bring general relativity and quantum field theory [64] together, physicists suggested that all matter must have a linear structure with 10 dimensions like Calabi-Yau manifold (Fig. E). This is known as the "String Theory" [4].

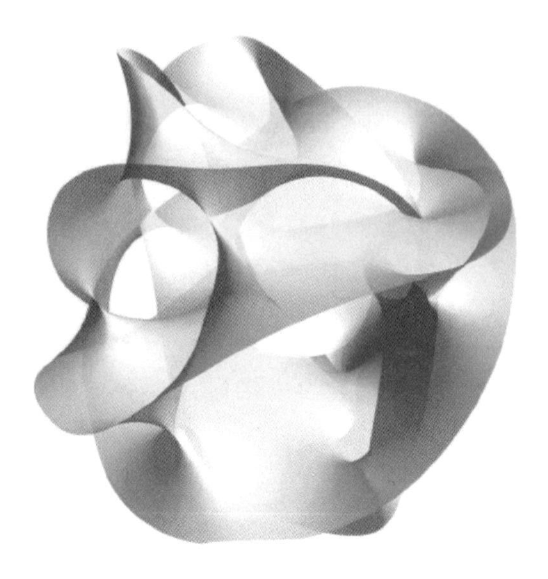

Fig. E A cross section of a quintic Calabi-Yau manifold.

Could String Theory be true? The answer is yes and only if all the subatomic particles have a linear structure. Physicists have absolutely no idea what the structures of quarks and photon are, even with their state-of-the-art LHC [5]. However, based on the Yangton and Yington Theory [1], that all subatomic particles should have a string structure is not only very possible, but also quite obvious.

Wu's Pair [1] is a pair of Yangton and Yington Antimatter particles circulating in an orbit held by the inter-attractive Force of Creation between the two particles. When two Wu's Pairs come together with the same circulation direction, they stack up on each other at a locked-in position, where Yangton of the first Wu's Pair is lined up to the Yington of the second one due to the attraction between Yangton and Yington particles. This induced force between the two Wu's Pairs is called "String Force". By reproduction of the above process, a string or ring or other related structures of Wu's Pairs called "String Structure" can thus be formed (Fig. 3) [1], which complies very well with the "String Theory".

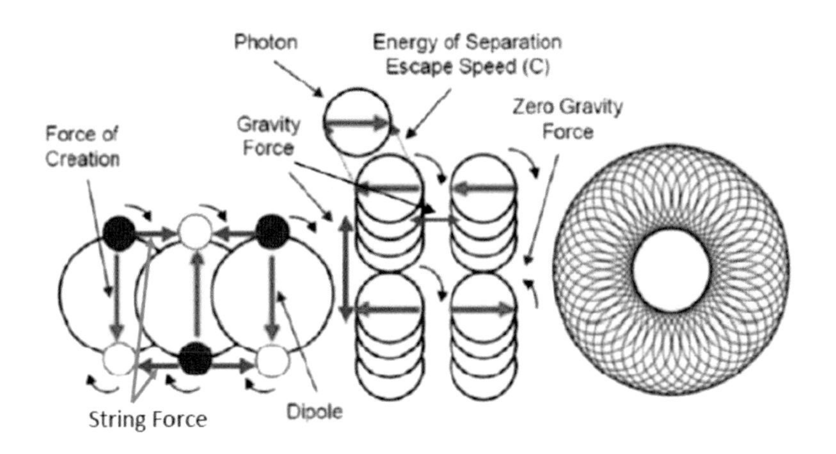

Fig. 3 Wu's Pairs stack up in a preferred direction by string force to form string and ring structures.

3.2. Subatomic Particles

Standard Model is a group of subatomic particles which is derived from a mathematical model based on quantum field theory and Yang Mills Theory. In contrast, Wu's Pairs, a physical model are proposed as the building blocks of all subatomic particles based on the Yangton and Yington Theory.

Subatomic particles [6] are very much smaller than atoms. There are two types of subatomic particles: elementary particles, which according to current theories are not made of other particles, and composite particles which are made of elementary particles. Particle physics and nuclear physics study these particles and how they interact.

The elementary particles of the Standard Model (Fig. 4) include:

- Six flavors of quarks: up, down, bottom, top, strange, and charm

- Six types of leptons: electron, electron neutrino, muon, muon neutrino, tau, tau neutrino

- Twelve Gauge Bosons (force carriers):
  the photon of electromagnetism, the three W and Z Bosons of the weak force, and the eight gluons of the strong force

- The Higgs Boson

Various extensions of the Standard Model predict the existence of an elementary graviton particle and many other elementary particles.

Composite subatomic particles such as protons or atomic nuclei are bound states of two or more elementary particles. For example, a proton is made of two up quarks and one down quark, a neutron is made of two down quarks and one up quark, while the atomic nucleus of Helium-4 is composed of two protons and two neutrons.

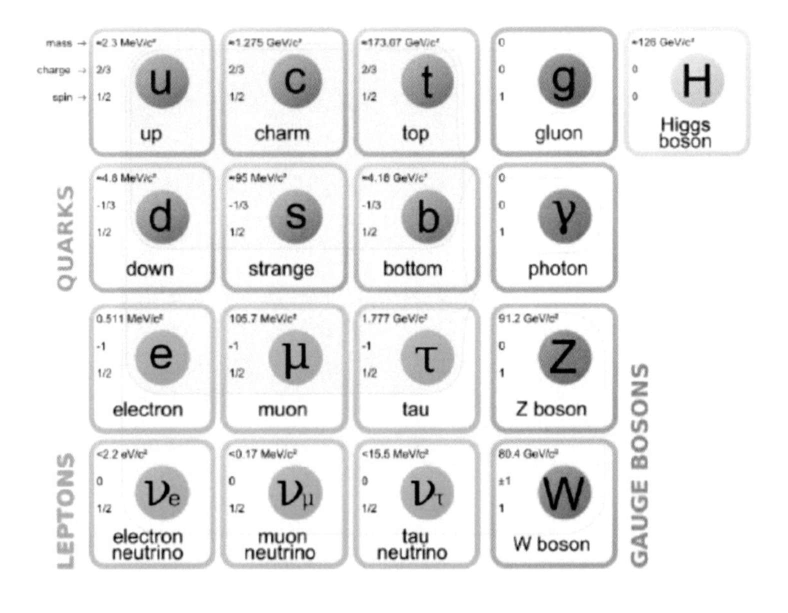

Fig. 4 The elementary particles of the Standard Model.

According to Yangton and Yington Theory, all elementary subatomic particles including quarks, leptons, Gauge Bosons, gluons and photon are made of Wu's Pairs. They have string, ring and other structures (Fig. 5) that are glued together by the string force between two adjacent Wu's Pairs (Fig. 3). Composite subatomic particles are made of elementary subatomic particles, which are glued together by four basic forces including gravitational force, electromagnetic force, weak force and strong force that are induced from Force of Creation subject to the subatomic structures and their interactions.

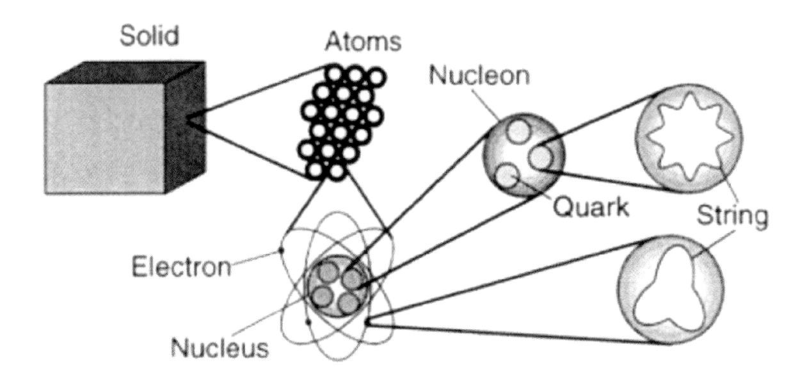

Fig. 5 Subatomic particles made of string structures.

3.3. Graviton and Gravitational Force

Wu's Pairs can be used to form elementary subatomic particles of string structures in a variety of shapes. When two string structures come together in the same circulation direction, they can attract each other at the ends of the strings by locking in the Yangton of one string to the Yington of the other string. Otherwise, there is no interaction if they are in the opposite circulation directions. However, when two string structures come together side by side, no matter the circulation directions, they can adjust themselves to attract each other as the Yangtons of one string contact the Yingtons of the other string during each cycle of the circulations. These attractive only forces are known as "Gravitational Force" (Fig. 6) and the string structures that produce the gravitational force are called "Gravitons".

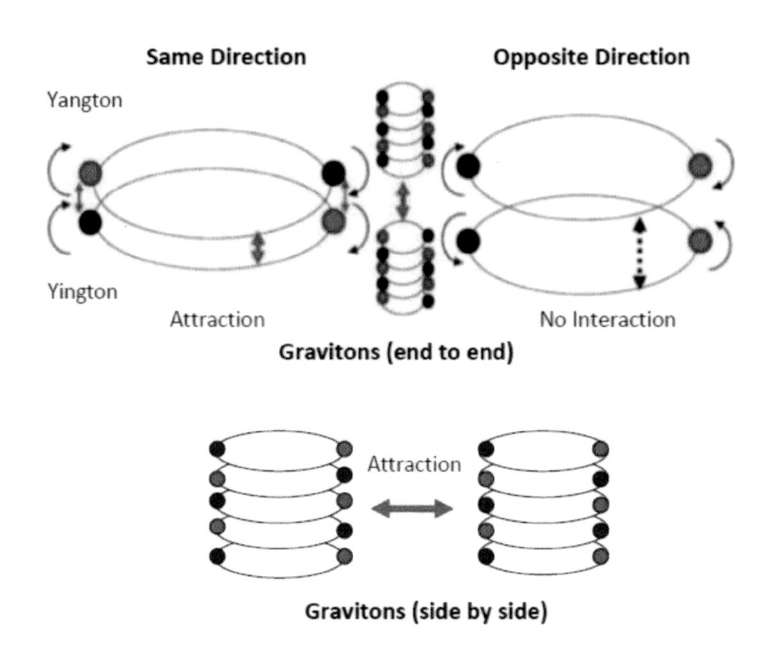

Fig. 6 Gravitational force between two graviton particles

3.4. Higgs Boson and Higgs Field

According to Standard Model and Quantum Field Theory, the mass of a particle is the magnitude of the barrier applied to the particle by "Higgs Bosons" that are generated from Higgs Field [7]. Since Higgs Bosons can be considered as the carriers of string force that are generated by Wu's Pairs, therefore the magnitude of the barrier caused by the string force carried by Higgs Bosons is proportional to the amount of Wu's Pairs. In other words, the mass of a particle is proportional to the amount of Higgs Bosons as is that of Wu's Pairs [75]. This concurs with that the mass is the total amount of Wu's Pairs based on Yangton and Yington Theory.

3.5. Electron, Positron and Electrical Force

When a number of Wu's Pairs come together they can stack up to form a string or ring structures (Fig. 3), or cross each other's orbits to form a structure that is either with Yingtons circulating the Yangton center as the electrons (Fig. 7) [2] or with Yangtons circulating the Yington center as the positrons (Fig. 7) [2].

Since photon, a free Wu's Pair, can be absorbed and emitted from an electron jumping between two energy levels in an atom; it is proposed that electron is composed of a group of Wu's Pairs, where Yangtons are loosely confined in the center due to the compression of the centrifugal force caused by the circulation of Yingtons. Therefore, electron can have an appearance looks like a sphere of Yingtons, and positron can have the appearance looks like a sphere of Yangtons (Fig. 7).

Because of the attraction between Yangton and Yington, a strong attractive force can be generated between an electron and a positron. Also, a repulsive force can be formed between two electrons as well as between two positrons. When a positron meets an electron, because of the attraction, they collide and destroy each other to release Gamma Ray (γ). This phenomenon is known as "Positron-Electron Annihilation" [63].

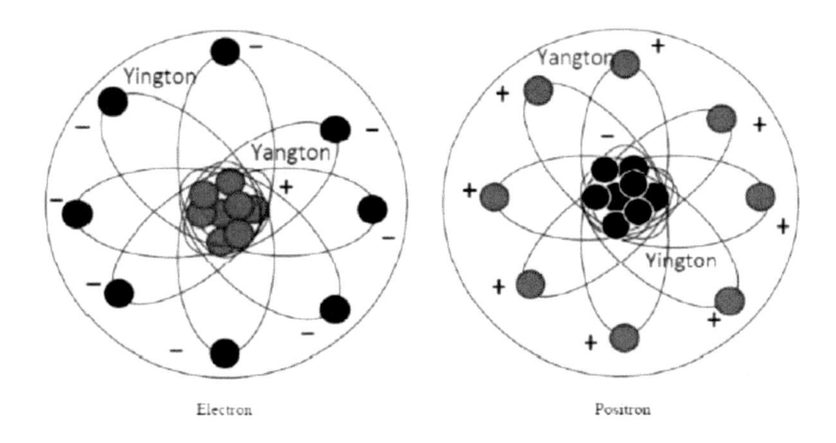

Fig. 7 Hypothetical structures of electrons and positrons.

How do I know about the electron and positron structures? The answer is, of course based on logical thinking:

1. The attractive force between Yangton and Yington is similar to that between electron and positron.

2. The repulsive forces between two Yangtons and that between two Yingtons are similar to that between two electrons and that between two positrons.

3. Since Wu's pairs are the building blocks, then electrons must be a cluster of Yingtons and positrons must be a cluster of Yangtons.

4. Since Wu's pair is made of a Yangton and Yington pair and they can not be separated from each other, therefore an electron can only be structured by a sphere made of Yingtons with a center made of Yangtons. Similarly, a positron can only be structured by a sphere made of Yangtons with a center made of Yingtons.

3.6. Proton, Neutron, Weak Force and Strong Force

A neutron [8] is composed of three quarks, one up quark and two down quarks, and three gluons. Since all matter have string structures of Wu's Pairs, it is believed that a neutron containing three quarks and three gluons should have the shape as a donut or a triangular pretzel (Fig. 8).

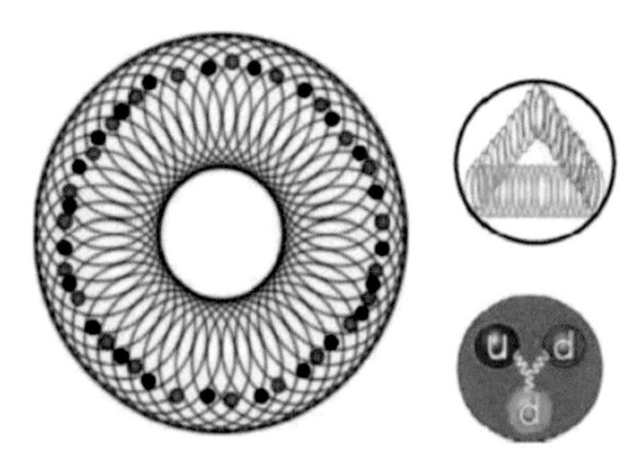

Fig. 8 A hypothetical structure of neutron.

A proton [9] is also composed of three quarks, two up quarks and one down quark, and three gluons. Therefore, like the neutron, a proton containing three quarks and three gluons should also have the shape as a donut or a triangular pretzel. However, because of the Inverse Beta Decay, it is believed that a proton contains a neutron with an embedded positron and electron neutrino (Fig. 9).

Fig. 9 A hypothetical structure of proton.

The bonding force between a neutron and a positron is known as "Weak Force" (Fig. 9) [10] which is induced by the multiple Yangtons on the surface of the positron.

In order to balance the repulsive electromagnetic force caused between protons, strong force is needed to hold protons together in the nucleus. Strong force is the attractive force generated between two neutrons, and also between a neutron and a proton. When two neutrons with ring structures made of Wu's Pairs come together, attractive force can be generated between the two neutrons with either the same or opposite circulation directions. This attractive force is known as "Strong Force" (Fig. 10) [11], which are many magnitudes larger than the gravitational force.

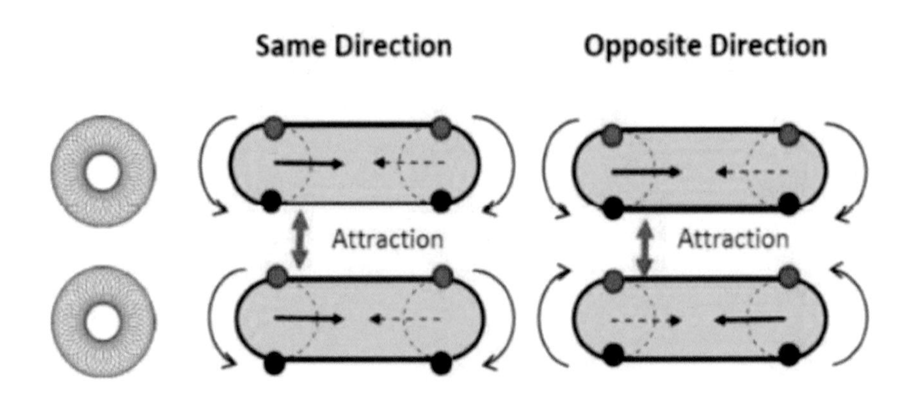

Fig. 10 Strong force between two neutrons.

When a neutron comes close to a proton made of a neutron, positron and electron antineutrino, both the weak force between neutron and positron (Fig. 11), and the strong force between neutron and neutron (Fig. 11) are generated to overcome the repulsive force between protons so as to keep them together inside the nucleus.

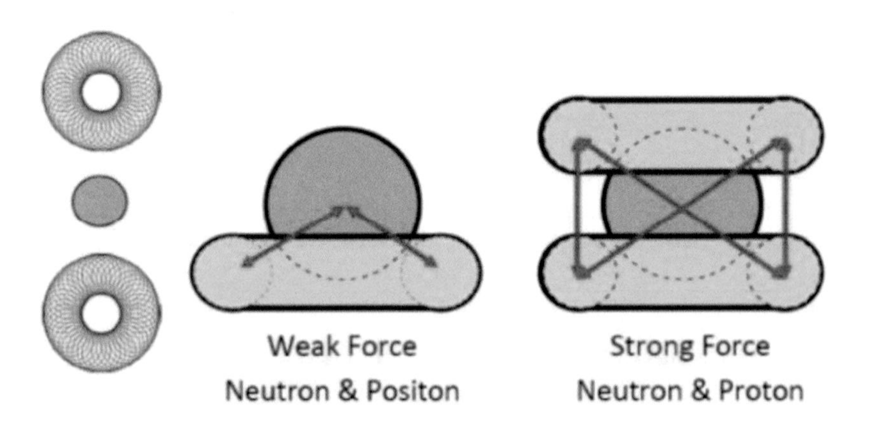

Fig. 11 Weak force and strong force.

3.7. Beta Decay and Inverse Beta Decay

Since positron is the Antimatter of electron and electron neutrino is the Antimatter of electron antineutrino, free neutron in Beta Decay at high energy state first incubates two pairs of Antimatter, an electron/positron pair and an electron neutrino/electron antineutrino pair, then emits the electron and electron antineutrino, while maintaining the neutron, positron and electron neutrino to form a proton at low energy state. This phenomenon is known as "Beta Decay".

$$n^0 = n^0 + (e^+ + e^-) + (\nu_e + \underline{\nu}_e) = (n^0 + e^+ + \nu_e) + e^- + \underline{\nu}_e = p^+ + e^- + \underline{\nu}_e$$

$$n^0 \rightarrow p^+ + e^- + \underline{\nu}_e \text{ (Beta Decay)}$$

In Inverse Beta Decay, the weak force between the neutron and positron in a proton is overcome by the kinetic energy of the proton such that a proton can transfer to a neutron by emitting a positron and an electron neutrino.

$$p^+ \rightarrow n^0 + e^+ + \nu_e \text{ (Inverse Beta Decay)}$$

3.8. Antimatter and Baryogenesis

Antimatter [12] is a material composed of antiparticles, which have the same mass as particles of ordinary matter but opposite charges, as well as other particle properties such as Lepton and Baryon Numbers. Collisions between particles and antiparticles leads to the annihilation of both, giving rise to variable proportions of intense photons (gamma ays), neutrinos, and less massive particle–antiparticle pairs. The total consequence of annihilation is a release of energy available for work, proportional to the total matter and Antimatter mass.

There is considerable speculation as to why the observable universe is composed almost entirely of ordinary matter, but an even mixture of matter and Antimatter. This asymmetry of matter and Antimatter in the visible universe is one of the great unsolved problems in physics. The process by which this inequality between particles and antiparticles developed is called "Baryogenesis".

According to Yangton and Yington Theory, Yangton is the Antimatter of Yington, such that they can destroy each other and release energy during the recombination. Most subatomic particles composed of Wu's Pairs have symmetrical structures which inhibits the formation of their Antimatters. Only a few polarized particles such as electrons and positrons (Fig. 7) having asymmetrical distribution of Yangton and Yington particles, their Antimatters could be formed. This explains well the Baryogenesis.

3.9. Dark Matter

Dark Matter [13] like any other subatomic particle is composed of a number of Wu's Pairs. It is proposed that Dark Matter has a tetrahedral structure of four Yangton and Yington Pairs (Fig. 12). Each Yangton and Yington Pair is circulating on its own orbit at 109.5° away from the other three pairs. Because the Yangton center coincides to the Yington center, there is no dipole in the center of the tetrahedral structure. As a result, there is no attractive force between two Dark Matters neither between Dark Matter and any other substances. In other words, there is no gravitational force can be generated by Dark Matter in any way.

Because there is no attractive force, Dark Matter cannot be used as the building block of any substance. Also, because it is very stable and there is no particle can escape from it, Dark Matter is totally invisible and that is why it is called "Dark Matter".

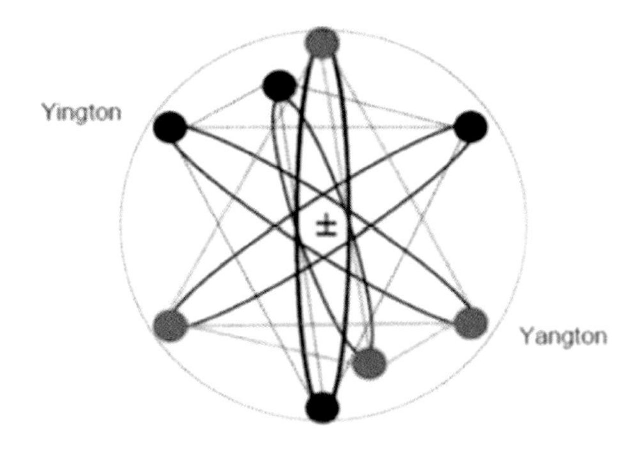

Fig. 12 A hypothetical tetrahedral structure of Dark Matter.

3.10. Electrical Force and Magnetic Force

In addition to the attractive and repulsive electric forces formed between the static electrons, positrons and protons, a secondary magnetic force can be generated between the moving electrons, positrons and protons. When two atoms, each carrying a single outer layer electron, come together with their outer layer electrons circulating and spinning in the same direction, they prefer to stay 180 degrees away from each other so as to form an attractive force between the two atoms. Otherwise, the repulsive force will be formed with their outer layer electrons circulating and spinning in the opposite directions (Fig. 13). This is known as "Magnetic Force" [14].

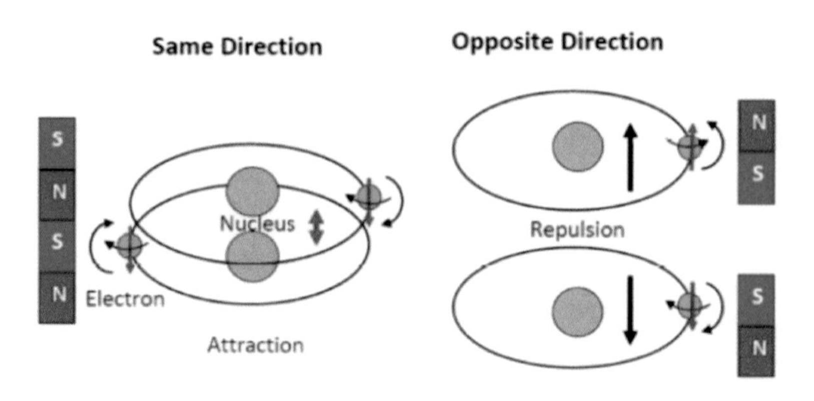

Fig. 13 Electromagnetic force between two atoms each with a single outer layer electron.

3.11. Unified Field Theory

According to Yangton and Yington Theory, Force of Creation, the inter-attractive force between Yangton and Yington Pairs is the fundamental force of the universe. Elementary subatomic particles of string structure are composed of Wu's Pairs with string force generated from Force of Creation. Composite subatomic particles are made of elementary subatomic particles with four basic forces also induced from Force of Creation. In other words, subject to the structures, all matter in the universe can be composed of elementary and composite subatomic particles with four basic forces based on Force of Creation. This is known as "Unified Field Theory" [15]. For example, based on Force of Creation, gravitational force can be created between two Graviton particles, electromagnetic force can be generated between electrons and protons, weak force can be formed between neutrons and positrons, and strong force can be produced between two neutrons, also between a neutron and a proton.

In the past few decades, physicists tried to develop a quantum gravity theory in accompany with quantum field theories as a unified field theory to explain the formation of four basic forces.

So far, there has been a little success due to the incompatibility between general relativity and quantum field theory. However, based on Wu's Pairs and Yangton and Yington Theory, subatomic particles with string structures made of Wu's Pairs and Force of Creation can easily comply with unified field theory.

3.12. Wave Particle Duality and De Broglie Matter Wave

It is believed that only spinning particles such as photon and electron can have Wave Particle Duality. Since graviton doesn't spin therefore it doesn't have Wave Particle Duality.

A spinning particle, like a whirlpool, has a momentum that is proportional to the mass m and the spin frequency ν [74]:

$$P = Km\nu$$

$$P = KmC/\lambda$$

Where K is whirlpool constant.

Therefore, the energy and momentum of a spinning particle can be calculated as follows:

Because

$\Delta E = P\ \Delta V$

Assuming the final speed of the particle is Absolute Light Speed C, then

$E = PC = Km\nu C$

Given

$$h_0 = KC$$

Therefore,

$$E = mh_0\nu$$

$$P = mh_0/\lambda$$

And the wavelength of the spinning particle is

$$\lambda = mh_0/P$$

Where m is the mass, E is the energy and P is the momentum of the spinning particle. h_0 is the General Planck Constant and λ is the wavelength of De Broglie Matter Wave of a spinning particle.

A. Photon

The momentum of a photon, similar to a whirlpool, is proportional to the mass of the photon (m_{yy}) and the frequency (ν) of the circulation of Wu's Pair. Therefore,

$$P = Km_{yy}\,\nu$$

Where K is whirlpool constant and m_{yy} is the mass of a photon or a single Wu's Pair.

Since energy is the amount of interaction applied on an object caused by the resistance of the action and the response of the action, therefore, the energy difference (ΔE) between a Wu's Pair and a photon can be represented by the multiplication of the momentum (P) and the change of velocity (ΔV) as follows:

$$\Delta E = P\Delta V$$

Because,

$\Delta E = E$

Also,

$\Delta V = C$

Therefore,

$$E = PC$$

Where E is the kinetic energy, P is the momentum of a photon and C is the Absolute Light Speed.

Because,

$P = Km_{yy}\nu$

$$h_0 = KC$$

Therefore,

$E = Km_{yy}\nu C = (m_{yy}KC)\nu = m_{yy}h_0\nu$

Given

$$h = m_{yy}h_0$$

Therefore,

$$E = h\nu$$

Where h is Planck constant 6.626×10^{-34} m^2 kg/s. E is the kinetic energy and ν is the frequency of a photon.

Because,

$P = E/C = (h\nu/C) = h/(C/\nu)$

$C/\nu = \lambda$

Therefore,

$$P = h/\lambda$$

And

$$\lambda = h/P$$

Where P is the momentum and λ is the wavelength of a photon and h is Planck constant 6.626×10^{-34} m^2 kg/s.

It is believed that during the photon separation process, the string energy associated with the string force between two Wu's Pairs on the surface of the substance is converted to the kinetic energy of the photon.

B. Electron

For an electron circulating around the atomic nucleus, the wavelength of the electron can be represented as:

$$\lambda = m_e h_0 / P$$

Where m_e is the mass of an electron and h_0 is the General Planck Constant.

According to Bohr model,

$2\pi R = n\lambda = n m_e h_0 / P$

$m_e(V^2/R) = KZ(e^2/R^2)$

$E = -\frac{1}{2} m_e V^2$

Also,

$h = m_{yy} h_0$

$\hbar = h/2\pi$

Therefore,

$$\lambda = (m_e/m_{yy})h/P$$

$$V = (KZe^2/m_e R)^{1/2}$$

$$R = (n^2\hbar^2/m_e KZe^2)(m_e/m_{yy})^2$$

$$P = m_e(KZe^2)/(n\hbar)(m_e/m_{yy})$$

$$E = -\frac{1}{2} m_e(KZe^2)^2/(n^2\hbar^2)(m_e/m_{yy})^2$$

Since m_e/m_{yy} is the amount of Wu's Pairs in an electron, it is a constant and so is $(m_e/m_{yy})^2$.

Spinning particles are simulated by a whirlpool model. Because the momentum P of the spinning particle is proportional to the mass m and the spin frequency ν of the particle, therefore $P = Km\nu = KmC/\lambda = mh_0/\lambda$ (where h_0 is General Planck Constant), De Broglie Wavelength $\lambda = mh_0/P$ and Planck constant $h = m_{yy}h_0$ (where m_{yy} is the mass of a photon or a Wu's Pair) [74].

Furthermore, De Broglie wavelength, momentum and energy of the electron in Bohr Model are calculated. As a result, all Planck constant h in the old formula are replaced by $(m_e/m_{yy})h$ in the new version [74].

3.13. Uncertainty Principle

Introduced first in 1927, by the German physicist Werner Heisenberg, the Uncertainty Principle states that the more precisely the position of some particle is determined, the less precisely its momentum (or velocity) can be predicted from initial conditions, and vice versa [77]. The formal inequality relating the standard deviation of position ΔX and the standard deviation of momentum ΔP was derived by Earle Hesse Kennard later that year and by Hermann Weyl in 1928 as:

$$\Delta P \, \Delta X \geq \hbar /2$$

This can also be represented as:

$$\Delta E \, \Delta t \geq \hbar /2$$

Where ΔP is the standard deviation of momentum, ΔX is the standard deviation of position, ΔE is the standard deviation of energy, Δt is the standard deviation of time and ħ is the reduced Planck constant $h/(2\pi)$.

Uncertainty Principle plays an important roll in Subatomic Particle Interactions such as Quantum Electrodynamics (QED) and Quantum Chromodynamics (QCD).

3.14. Wu's Pairs versus Standard Model

According to Yangton and Yington Theory, all subatomic particles are made of Wu's Pairs, therefore in theory Standard Model and quantum field theory should align with Wu's Pairs and Yangton and Yington Theory [60]. However, there are some differences that are addressed as follows:

3.14.1. Special Relativity

Standard Model and quantum field theory are derived based on quantum mechanics and Einstein's Special Relativity [39]. Although special relativity is opposed by Yangton and Yington Theory [34], the absolute light speed observed at light source is always a constant 3×10^8 m/s, simply because that the photon generation process [28] is a corresponding identical event, where constant light speed can be measured by the corresponding units at the reference points. Also, since Wu's Pairs are the finest building blocks of all matter and the photon separation force, the string force between two Wu's Pairs, is the strongest force between two objects except Force of Creation, it is suggested that the Absolute Light Speed 3×10^8 m/s is the limit of the speed that any object can move in the universe.

However, the main reason that Special Relativity is adopted in Standard Model and quantum field theory is because of the relativism, otherwise the energy of photon and gluons will be zero (both have zero mass).

3.14.2. Mass and Energy Conversion

It is assumed that mass and energy are convertible such as that Yangton and Yington particles (could also be considered as energy particles) can be produced by the energy generated from the Big Bang explosion.

$$E_{Creation} + E_{Circulation} \leftrightarrow \text{Yangton } \Phi \text{ Yington}$$

The conversion between Mass and Energy can also be commonly found in LHC experiments [5]. In some cases, an external energy must be applied to overcome the activation energy like that in chemical reactions. Because of these reasons, antiparticle pairs can be formed from vacuum by external energy. Heavy particles can be produced from light particles [10] and gamma ray can be generated from antiparticle annihilations [63]. Furthermore, a virtual photon [64] can be used to represent an energy transformation process.

3.14.3. Symmetry

A. Anti Particles

According to Yangton and Yington Theory, all subatomic particles are made of Wu's Pairs, a pair of Yangton and Yington circulating Antimatter particles. Because of the symmetry of Yangton and Yington in the circulation, it makes no difference in the structure while switching the positions between Yangton and Yington particles. Therefore, antiparticles such as anti-photon and anti-gluons do not exist. This explains the reasons of Baryogenesis.

However, for a structure made of uneven distributions of Yangton and Yington in Wu's pairs such as electron (Fig. 7), a different structure can be formed while switching the positions between Yangton and Yington particles. Therefore, antiparticle such as positron (anti-electron) (Fig. 7) can be formed.

B. Colors of Quarks and Gluons

In addition to the asymmetry due to the uneven distribution of Yangton and Yington in Wu's Pairs, there is another asymmetry called "Color" [6] specified by red, blue and green colors that are related to the orientation between two connected quarks. Because each proton or neutron contains three quarks and each quark only allows one color (red, blue or green), there are only eight gluons with the following arrangements: UDU/RBG,

UDU/RGB, UUD/RBG, UUD/RGB, DDU/RBG, DDU/RGB, DUD/RBG and DUD/RGB. For example, UDU/RBG represents a gluon connected between two up quarks with red and green colors, influenced by a down quark of blue color.

C. Strong Forces Between Proton/Neutron and Neutron/Neutron Pairs

According to Yangton and Yington Theory, in a proton or neutron, gluons [6] (the strong force carriers) are the connectors of two quarks influenced by the third quark with a mixed color of preferred orientation. But what is the bonding force between two neutrons or between one neutron and one proton in the nucleus? It is proposed that the gluon in one neutron will comply with its influence quark in the other neutron or proton to form a close packed structure (Fig. F). Therefore, the force between two adjacent neutrons, or a neutron and proton pair has no difference to the strong force between the three quarks inside a single neutron or proton.

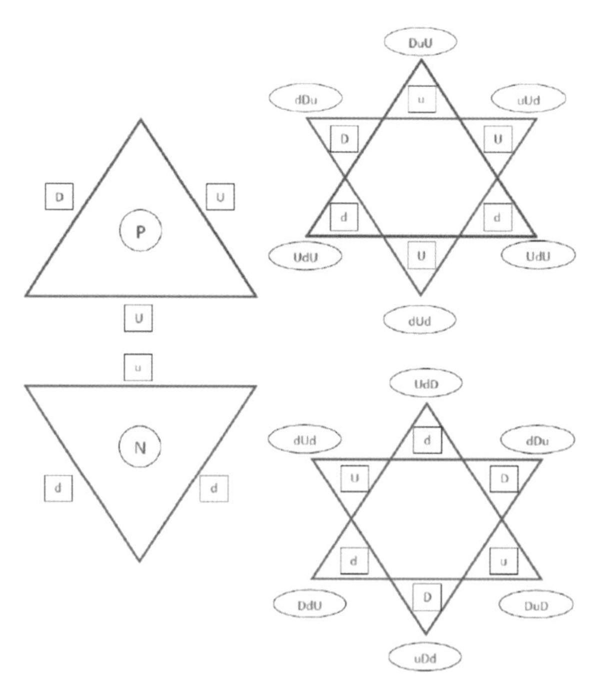

Fig. F Strong Forces between Neutron/Proton and Neutron/Neutron Pairs

3.14.4. Gauge Bosons

In Standard Model [6], Gauge Bosons including photon, gluons, W & Z Bosons and graviton are considered as force particles or force carriers. However, according to Yangton and Yington Theory, Photon like Wu's Pair is the carrier of string force instead of electromagnetic force. Also, electromagnetic force is generated by electrons, therefore electron should be considered as the carrier of electromagnetic force instead of photon.

All Gauge Bosons are force carriers, like photon shouldn't contain any charge and mass. W^+ and W^- Gauge Bosons, however, both have charges and masses because they are not made of pure force carriers, instead, they are the composites of forces, particles (mass) and charges (electron and positron). Zero Gauge Boson and graviton on the other hand contain forces and particles (mass) but charges.

3.14.5. Higgs Boson, Higgs Field and Mass

In Standard Model and quantum field theory, the mass of a particle is the magnitude of the barrier applied to the particle by Higgs Bosons that are generated from Higgs Field. However, what the Higgs Bosons and Higgs Field are and where they come from remain mysteries. According to Yangton and Yington Theory, mass is the total amount of Wu's Pairs [28] in a particle which are bound together by string force. Therefore, a Higgs Boson can be considered as a particle which contains two Wu's Pairs bound together by string force, and Higgs Field can be interpreted as the distribution of Higgs Bosons in the space as is that of string force [75].

3.14.6. Field versus Wave

Wave can be made of a spin particle, but not field. Field reflects the distribution of a physical quantity in space such as particle concentration. In the case of gravitational field, it reflects the distribution of particle contact interactions (gravitational force) that is proportional to particle radiation concentration. This is named Particle Radiation and Contact Interaction Theory [17]. This explains why photon is a wave (with spin) but not a field (without contact interaction), and graviton is a field (with particle radiation and contact interaction) but not a wave (without spin and circulation).

Wu's Pairs and Higgs Bosons (String Force Carriers) both spin by themselves, therefore they can be considered as a wave. In addition, interactions can be generated by the string forces between Wu's Pairs and Higgs Bosons, which makes them a field. This matches very well with Higgs Field Theory [75].

3.14.7. Quantum Field Theory

According to Newton's Law of Universal Gravitation, gravitational field is the measurement of all the gravitational force in the universe applied on a unit mass at a point in space. Based on Yangton and Yington Theory, it is equivalent to the total amount of contact interaction caused by the gravitons emitted from all the objects in the universe upon a unit mass at a point in space. In other words, gravitational field is the distribution of the graviton concentration vectors caused by graviton radiation from all the objects in the universe [75].

Similarly, the electrical field which is defined as the electrical force applied on a single electrical charge in the universe can be interpreted as the distribution of the electron concentration vectors caused by electron radiation from all the charged particles in the universe. Magnetic field on the otherhand is an induced field from an electrical field. Together they form an electromagnetic field.

As a result, with Particle Radiation and Contact Interaction Theory [17], the distribution of concentration vectors of the free particles caused by particle radiation from all the objects in the universe can be considered as the foundation of quantum field theory. In other words, quantum field is not a pure imagination; instead, it is a field that is quantized by the distribution of concentration vectors of the free particles caused by particle radiation in the universe [75].

3.14.8. Quantum Gravity Theory

There are two problems, infinity and non-renormalization [61], involved in the derivation of a quantum gravity theory based on general relativity and quantum field theory. To avoid the problems, string theory [4] and loop quantum gravity [62] are proposed as two possible solutions.

According to Yangton and Yington Theory, gravitational force can be generated between two gravitons with string structures made of Wu's Pairs. In addition, quantum field theory is considered as the distribution of concentration vectors of the free particles caused by particle radiation in the universe. Therefore, it is believed that "Quantum Gravity Theory" can be derived based on the gravitons with a string structure made of Wu's Pairs and the theory of graviton radiation and contact interaction [17] [75].

3.14.9. Quantum Field Theory versus Yangton & Yington Theory

In general, a conceptual comparison between quantum field theory and Yangton and Yington Theory can be represented as follows:

A. Quantum Field Theory

Quantum Mechanics (QM) + Special Relativity (3×10^8 m/s) → Quantum Field Theory (QFT)

Quantum Field Theory (QFT) + Yang-Mills Theory → Standard Model

General Relativity + String Theory (Calabi-Yau Manifold + Quantum Field Theory (QFT)) → Quantum Gravity Theory (QGT) → Unified Field Theory (UFT)

B. Yangton & Yington Theory

Point Particle + Particle Radiation & Contact Interaction (PRCI) → Quantum Field Theory

String Particle + Particle Radiation & Contact Interaction (PRCI) → Quantum Gravity Theory (QGT) → Unified Field Theory (UFT)

In general, Wu's Pairs and Yangton and Yington Theory provide a particle model and physical picture for the interpretation of Quantum Field Theory, Quantum Gravity Theory and Unified Field Theory.

Chapter Four

Propagation of Force

Does Gravitational Force Propagate?

4.1. Graviton Radiation and Contact Interaction versus Newton's Law of Universal Gravitation

Newton's Law of Universal Gravitation [16] only describes the phenomenon of the gravitational force without explaining what the process is and how it works. Particle Radiation and Contact Interaction Theory [17] is proposed to explain the mechanism and the propagation of the gravitational force.

According to Yangton and Yington Theory, gravitational force is generated between two gravitons at close contact as shown in Fig. 3 and Fig. 6. Because of this reason, for two distance objects, a graviton particle must first escape from the parent object, and then travel to the target object to make a contact, so that the propagation of gravitational force can be accomplished.

Like a photon emitted from a heat source by absorbing thermal energy to overcome its energy of separation, graviton can also be emitted from an object by absorbing thermal energy to overcome its gravitational force. It is obvious that the concentration of the emitted gravitons should be proportional to the total amount of gravitons in the parent object, which is the mass of the parent object (m_1). According to the Inverse Square Law [18], the concentration of the emitted gravitons that reaches the target object should be inversely proportional to the square of the distance (r) between parent object and target object. Furthermore, because of the random orientations from 0° to 90° between the emitted graviton and the graviton on the target surface (Fig. 14) [17], in average only 50% of the full contact interactions can be generated from the target object (m_2).

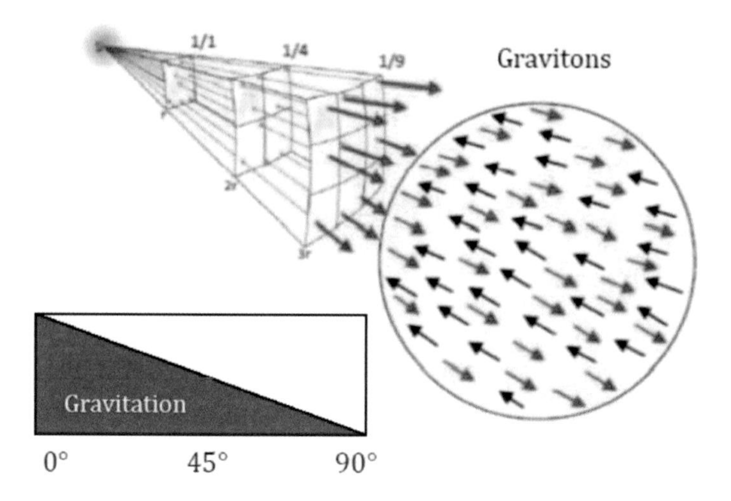

Fig. 14 Gravitational force caused by Graviton Radiation and Contact Interaction.

Therefore, the total gravitational force generated at any time should be proportional to the concentration of the emitted gravitons from the parent object that hit the surface of the target object (m_1/r^2) and the total amount of gravitons on the target object (m_2). A formula such as Newton's Law of Universal Gravitation (Fig. 15) can thus be derived as follows:

$$F = G\,(m_1 m_2 / r^2)$$

Where G is the gravitational constant 6.674×10^{11} N m^2 kg^{-2} [Annex 30].

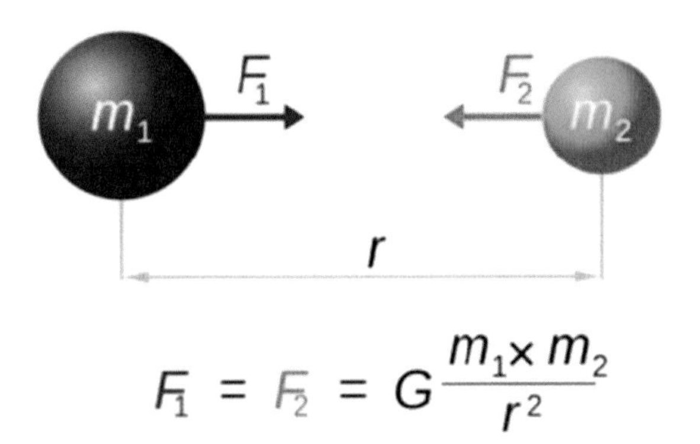

Fig. 15 Gravitational force between two objects.

4.2. Electron Radiation and Contact Interaction versus Coulomb's Law of Electrical Force

Similar to the gravitational force between two objects, the electric force between two charged particles should be proportional to the amount of charge in parent charged particle (q_1) and target charged particle (q_2) and inversely proportional to the square of the distance (r) between the parent charged particle and the target charged particle. Therefore, a formula such as Coulomb's Law of Electrical Force (Fig. 16) [19] can be derived as follows:

$$|F| = k_e\,(|q_1\,q_2|\,r^{-2})$$

Where k_e is Coulomb's constant 8.99×10^9 N m^2 C^{-2}. If the product q_1q_2 is positive, the force between the two charges is repulsive. If the product is negative, the force between them is attractive.

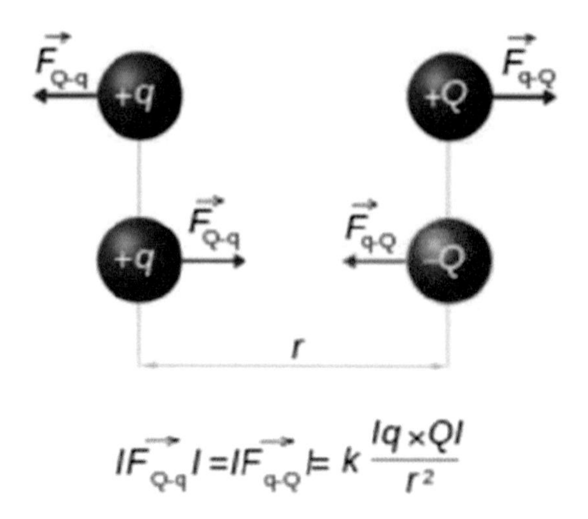

Fig. 16 Electrical force between two charged particles.

Electron radiation from a negatively charged particle has a much smaller scale than that of the Graviton Radiation. Because not only the negative charges (free electrons) are repulsive to each other, but also they can be easily neutralized by the positive charges and trapped by electron holes.

4.3. Gravitational Field and Concentration of Gravitons

Gravitational field is the strength of gravitational force in the universe applied on an unit mass (1 Kg) at a point in space. According to Newton's Law of Universal Gravitation, a formula of gravitational field (F_g) can be derived as follows:

$$F_g = G\,(\sum m\, r^{-2})$$

Where G is the gravitational constant 6.674×10^{11} N m^2 kg^{-2} and $\sum$ is the summation of all m r^{-2}.

Since mr^{-2} is proportional to the concentration of the free gravitons emitted from an object m at a distance r, therefore $\sum m$ r^{-2} is the summation of concentration vectors of the free gravitons emitted from all the objects in the universe to a point in space. In other words, gravitational field is the distribution of concentration vectors of the free gravitons caused by graviton radiation from all the objects in the universe [17].

Similar to gravitational field, the electrical field which is defined as the strength of the electrical force applied on a point in space can thus be interpreted as the distribution of concentration vectors of the free electrons caused by electron radiation from all the charged particles in the universe.

As a result, both the gravitational and electrical fields can be explained by a quantum theory based on the distribution of concentration vectors of the free particles through Particle Radiation and Contact Interaction. In other words, this theory provides a particle model and physical picture for the interpretation of Quantum Field Theory, Quantum Gravity Theory and Unified Field Theory.

4.4. Gravitational Wave

Gravitational waves [20] are proposed as "ripples" in space-time that propagate like waves, traveling outward from the source caused by some of the most violent and energetic processes in the universe. Albert Einstein predicted the existence of gravitational waves in 1916 in his general theory of relativity [21]. Einstein's mathematics show that massive accelerating objects would disrupt space-time in such a way that "waves" of distorted space would radiate from the source. Furthermore, these "ripples" would travel at the speed of light through the universe, carrying with them information about their cataclysmic origins, as well as invaluable clues to the nature of gravity itself. By contrast, gravitational waves cannot exist in Newton's theory of gravitation, since Newton's theory postulates that physical interactions propagate at infinite speed. Potential sources of detectable gravitational waves include binary star systems composed of white dwarfs, neutron stars, stellar cores (supernovae) or Black Holes. On February 11, 2016, the LIGO Scientific Collaboration and Virgo Collaboration teams announced that they had made the first observation of gravitational waves, originating from a pair of merging black holes using the Advanced LIGO detectors [22].

Based on Particle Radiation and Contact Interaction Theory, gravitational force is delivered via particle radiation rather than wave propagation. Instead of the wave form, Gravitational Waves are actually the fluctuation of the total gravitational forces carried by particle radiation generated from two merging Black Holes. Like the total brightness (Fig. 17) caused by two circulating stars, Gravitational Waves are generated by one Black Hole blocking the Graviton Radiation of another Black Hole during their circulation, which is detected in line of sight by earth observers. However, unlike that of the brightness, the fluctuation of the gravitational force is very small. Only massive stars or Black Holes can block (or refract) the Graviton Radiations from the other stars or Black Holes. Therefore, the Gravitational Waves are hard to detect except those caused by the massive binary star rotation system or a pair of merging Black Holes.

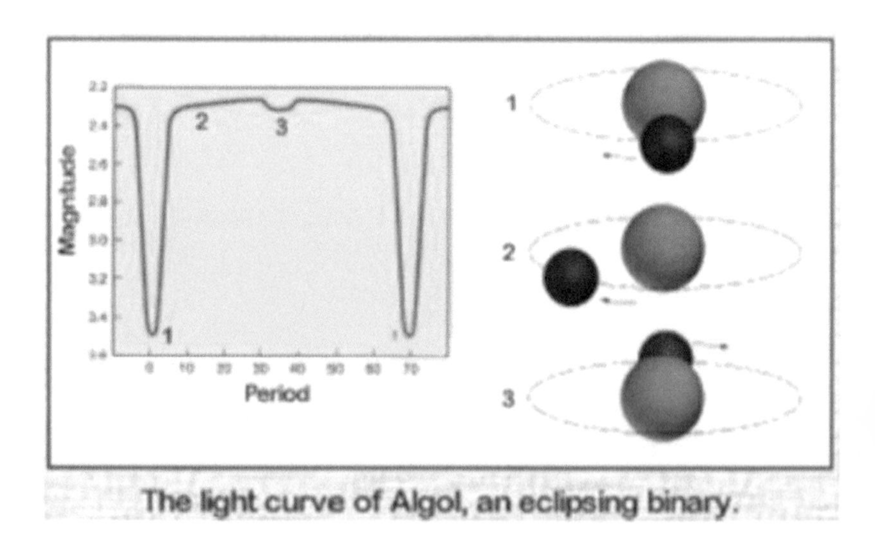

The light curve of Algol, an eclipsing binary.

Fig. 17 The brightness of the eclipsing “Algol” binary system.

Chapter Five

Mass and Force

What Is Mass?

Why F = ma?

5.1. Mass

Mass (M) is the total substance contained in an object. Because Wu's Pairs (Yangton and Yington circulating Antimatter particle pairs) are the building blocks of all matter, therefore mass (M) is the total amount of Wu's Pairs (m) times Wu's Unit Mass (m_{yy}), the mass of a single Wu's Pair (a constant, no matter of the gravitational field and aging of the universe).

Inertia is the measurement of the resistance of an object against motion. For an object with big inertia, larger force is needed to make the same acceleration. The inertia of an object is proportional to the mass of the object.

5.2. Momentum

"Momentum" (P) is the measurement of the moving ability of an object. For the same object, its momentum is proportional to its moving velocity (V). For different objects, their momentums at the same velocity are proportional to their masses (M), the total amount of Wu's Pairs (m) times Wu's Unit Mass (m_{yy}).

Therefore,

$$P = MV$$

Where P is the momentum, M is the mass and V is the velocity of a moving object.

5.3. Force

The external force (F) applied on an object is defined as a small change of the momentum (dP) of the object over a small change of the time duration (dt).

$$F = dP/dt$$

Therefore,

$$P = \Sigma dP = \Sigma(F \times dt)$$

5.4. Law of Conservation of Momentum

Momentum (P) can be transferred between two objects through the interactive force and the period of duration. In the transition process, one object is donor and the other one is the acceptor. The donor gives the momentum and the acceptor receives. During the momentum transition process, the velocity of the donor decreases with a negative external force, in comparison to that the velocity of the acceptor increases with a positive external force. Because the external forces applied on the two objects have the same value but opposite directions, and the time duration of the transition process are the same for each object, therefore the total momentum of the system remains the same. This is known as the "Law of Conservation of Momentum" [23].

$dP_{Acceptor} = F\ dt \qquad (dV_{Acceptor} > 0)$

$dP_{Donor} = (-F)\ dt \qquad (dV_{Donor} < 0)$

Therefore,

$$dP_{Acceptor} + dP_{Donor} = 0$$

5.5. Newton's Second Law of Motion

Because

$P = MV$

$dP = MdV$

M is a constant,

$dP = Fdt$

Therefore,

$F = (MdV)/dt = M (dV/dt)$

$$F = Ma$$

This is known as the "Newton's Second Law of Motion" [24].

5.6. Mass and Force

Newton's Second Law of Motion claims that the motion of an object can be influenced by force based on the following formula:

$F = M_N a$

Newton's Mass (M_N) is a quantity associated with each object, rather than the amount of Wu's Pairs (m) times Wu's Unit Mass (m_{yy}), or the total mass (M) of the object. Because Newton didn't know what the mass really was, Newton's Mass (M_N) is often inversely defined as:

$M_N = F/a$

However, based on Wu's Yangton & Yington Theory, mass (M) is the total amount of substance, also the total amount of Wu's Pairs (m) times Wu's Unit Mass (m_{yy}) in the object. Therefore,

$$F = Ma = m\ m_{yy}\ a = M_W\ a$$

Where F is the total force, M is the total mass of the object and M_W is Wu's Mass, the total amount of Wu's Pairs (m) times Wu's Unit Mass (m_{yy}).

Newton's Mass (M_N) is equal to Wu's Mass (M_W) only because that Wu's Pairs are the building blocks of all the matter in the universe.

An object may have either or both gravitational mass and non-gravitational mass subject to the response of the object to different forces such as gravitational force, electrical force, mechanical force, etc. For example, photon has non-gravitational mass comparing to electron and Dark Matter both have gravitational masses (gravitons). Therefore,

$F_{Total} = F_{Gravitation} + F_{Electric} + F_{Mechanic}$

$M_{Total} = M_{Gravitation} + M_{Non\text{-}gravitation}$

$F_{Gravitation} = M_{Gravitation} E_{Gravitation}$

$F_{Electric} = QE_{Electric}$

$$F_{Total} = M_{Total}\, a$$

Where $E_{Gravitation}$ is gravitational field, $E_{Electric}$ is electrical field, $M_{gravitation}$ is the gravitational mass of those Wu's Pairs in the object affected by gravitational force and Q is the charge of the object affected by electrical force.

5.7. Mass and Newton's Laws

One of the problems in Newton's theories is that the meanings of "Mass" are often confusing and sometimes mistaken.

In Newton's Law of Universal Gravitation [16]:

$$F = GM_1M_2/r^2$$

Where F is the gravitational force ($F_{Gravitation}$). M_1 and M_2 are the gravitational masses ($M_{Gravitation}$), which each is only a portion of its own total mass (M_{Total}).

In Newton's Second Law of Motion [24]:

$$F = Ma$$

Where F is the total force (F_{Total}). M is the total mass (M_{Total}) instead of the gravitational mass ($M_{Gravitation}$).

However, in reality, gravitational mass ($M_{Gravitaion}$) is often mistakenly used as the total mass (M_{Total}).

$F' = M_{Gravitaion}\ a$

$F_{Total} = M_{Total}\ a$

$M_{Gravitaion} < M_{Total}$

Therefore,

$F' < F_{Total}$

The total force (F_{Total}) applied on an object is actually bigger than the force (F') based on the gravitational mass ($M_{Gravitaion}$).

5.8. Electric Force and Mass

A charged particle contains not only an electrical charge, which is responsive to the electrical field, but also the gravitational mass responsive to the gravitational force. Because there is no non-gravitational mass in a charged particle, gravitational mass is the total mass, which carries the total inertia against motion.

$F_{Total} = F_{Gravitation} + F_{Electric}$

$M_{Total} = M_{Gravitation}$

Therefore,

$F_{Gravitation} + F_{Electric} = M_{Gravitation}\ a$

Chapter Six

Energy

Why Einstein's Law $E = MC^2$?

6.1. Force and Energy

Energy is the amount of interaction applied on an object caused by the resistance of action and the response of action. For a moving object, the resistance of motion of the object is momentum (P) and the response of the motion of the object is the change of velocity (dV). Therefore,

$$dE = PdV$$

Because

$P = MV$

$dP = MdV$

$dE = PdV = MVdV = VdP$

Therefore,

$$dE = VdP$$

Also,

$dE = VdP = (dX/dt)dP = FdX$

Therefore,

$$dE = FdX$$

A. Kinetic Energy

Kinetic energy is the energy in an object associated with its motion.

Because,

$dE = F\, dX = Ma\, dX = M\, (dV/dt)\, dX = MV\, dV$

Therefore,

$E - E_0 = ½\, M\, (V^2 - V_0^2)$

Given,

$E_0 = 0,\ V_0 = 0$

Therefore,

$$E_k = ½\, MV^2$$

Where E_k is kinetic energy.

Also,

$P = MV$

Therefore,

$$E_k = P^2/2M$$

Where E_k is kinetic energy and P is momentum.

B. Potential Energy

Potential energy is the energy in an object associated with its mass and position. In a gravitational field, the difference of the potential energy in an object between the final position (present) and the initial position (infinity) is equal to the work done by the external force (gravitational force due to Graviton Radiation and Contact Interaction) moving the object from the initial position to the final position.

Therefore,

$$dE = Fdx$$

$$\sum_{E_i}^{E_f} dE = \sum_{r_i}^{r_f} Fdx$$

$$\int_{E_i}^{E_f} dE = E_f - E_i$$

Also,

$$F = K_g(M_e M)r^{-2}$$

$$\int_{r_i}^{r_f} Fdx = \int_{r_i}^{r_f} K_g(M_e M)r^{-2} \quad (-dr)$$

$$= K_g\,((M_e M)/r_f - (M_e M)/r_i)$$

Therefore,

$$E = K_g M_e M\,(1/r)$$

Where M_e is the mass of earth and K_g is a constant 6.674×10^{-11} $Nm^2\,kg^{-2}$.

C. Energy Conversion

In a gravitational field, an object of mass (M) at distance (r) from earth can gain a kinetic energy that is converted due to the reduction of potential energy from infinity (∞) to present distance (r). Therefore,

$$\frac{1}{2} MV^2 = K_g\,(M_e M)/r$$

Where M_e is the mass of earth and $K_g = 6.674\times10^{-11}\ Nm^2\,kg^{-2}$.

6.2. Law of Conservation of Energy

Energy can be transferred between two objects through their interactive force and motion. One object is the donor and the other is the acceptor. The donor gives energy with a negative external force and the acceptor receives it with a positive external force. In the energy transformation process, for any object, if the direction of the external force is the same as that of the direction of the motion, then the object is the acceptor and the counter part is the donor. The total energy remains the same, which is known as the "Law of Conservation of Energy" [25].

$dE_{Acceptor} = F\ dX$

$dE_{Donor} = (-F)\ dX$

$$dE_{Acceptor} + dE_{Donor} = 0$$

6.3. Einstein's Law of Conservation of Mass and Energy $E = MC^2$

When a matter explodes it becomes a bundle of free photons escaping into the space at a constant speed of 3×10^8 m/s. A massive energy in the magnitude of MC^2 is released. This theory is proposed by Einstein [26]. The theory predicts that matter and energy is interchangeable. Additionally a huge amount of energy can be released through the transformation (nuclear reaction).

Because photon is a free Wu's Pair traveling in space according to Yangton and Yington Theory, it is assumed that, during the explosion, a group of subatomic particles with mass M were first escaped into the space having kinetic energy ½ MC^2 at a light speed 3×10^8 m/s. And subsequently, Wu's Pairs were separated from the subatomic particles to form photons. Since all Wu's pairs become photons, the amount of Wu's Pairs remains unchanged during the explosion. Therefore, $E = MC^2$ has nothing to do with the transformation between mass and energy. In fact, it is an energy conversion from subatomic particle's structure energy (generated from string force and four basic forces) and kinetic energy to photon's kinetic energy $Mh\nu$ [68].

In fact, mass and energy conversion can be represented by the reaction of the formation of Wu's Pair as follows:

$$E_{Creation} + E_{Circulation} \leftrightarrow \text{Yangton } \Phi \text{ Yington}$$

Where Wu's Pair Formation Energy ($E_{Creation} + E_{Circulation}$) contains the formation energy of both Yangton and Yington energy particles, and the energies needed to generate Force of Creation and Force of Circulation. It is much bigger than MC^2 and can only be generated from either Big Bang explosion or nuclear reaction.

Chapter Seven

Properties of Photons

Is a Photon a Particle or a Wave?

7.1. Electromagnetic Wave

Because of the inter-attractive Force of Creation between Yangton and Yington Pairs (Wu's Pairs), it is assumed that a Yangton carries one positive electric unit charge and a Yington carries one negative electric unit charge. Together they form an electric dipole. These electric unit charges are the basic units of the normal electric charges that are carried by electrons, positrons and protons, except in a much smaller scale. In a photon, resulting from the circulation of the Yangton and Yington Pair (the rotation of the electric dipole), electromagnetic wave (Fig. 18) [27] can be generated and carried to a great distance at light speed.

Electron is different from Yington. It is proposed as a cluster of Wu's Pairs, which carry one negative electric charge that is made of a fixed amount of negative electric unit charges. Positron is also different from Yangton in a similar way except that they are made of a fixed amount of positive electric unit charges. A photon can be considered as a neutral particle simply because that any external force induced by the Yangton can be neutralized completely by that generated from the Yington because of the extremely small circulation orbit.

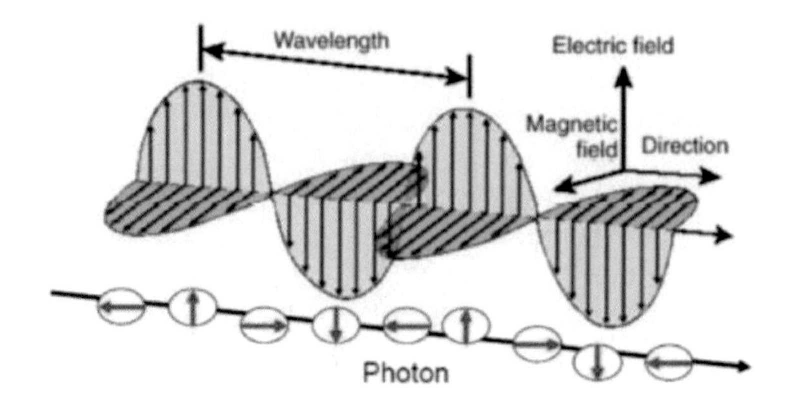

Fig. 18 Electromagnetic wave of a photon generated along traveling.

7.2. A Free Wu's Pair

A photon is a free Wu's Pair, a super fine Yangton and Yington circulating Antimatter particle pair, traveling at light speed in space. According to Yangton and Yington Theory, the mass of a photon is the same as that of a Wu's Pair, which is the mass of a pair of Yangton and Yington particles.

Since the circulation orbit is extremely small, any force induced by the Yangton (positive electric unit charge) can be neutralized by its counter force induced by the Yington (negative electric unit charge). In other words, photons can't be interfered with by any gravitational force or electromagnetic force. A photon has zero gravitational force on the surface of earth and sun, except in an extremely high gravitational field such as a Black Hole.

7.3. Photon Emission

A photon can be emitted from the parent object through a two stage process: separation stage and ejection stage (Fig. 19) [28].

A. Separation Stage

To unlock a photon from the surface of an object, it requires thermal energy to overcome the string force between the photon and the Wu's Pair on the surface of the parent object. Like a whirlpool, the momentum of a photon with a fixed mass (m_{yy}) is proportional to the frequency (ν) of the circulation of Yangton and Yington particle pair.

Because

$$P = Km\nu$$

For a photon,

$$P = Km_{yy}\nu$$

Given $K_1 = Km_{yy}$

$$P = K_1 \nu$$

Where K is whirlpool constant, m_{yy} is the mass of photon and K_1 is a constant associated with photon.

B. Ejection Stage

After the separation, photon is ejected by the repulsive forces between the two Yangton particles (one from the emitting photon and the other one from the adjacent Wu's Pair) also that between the two Yington particles (one from the emitting photon and the other one from the adjacent Wu's Pair) on the surface of the object.

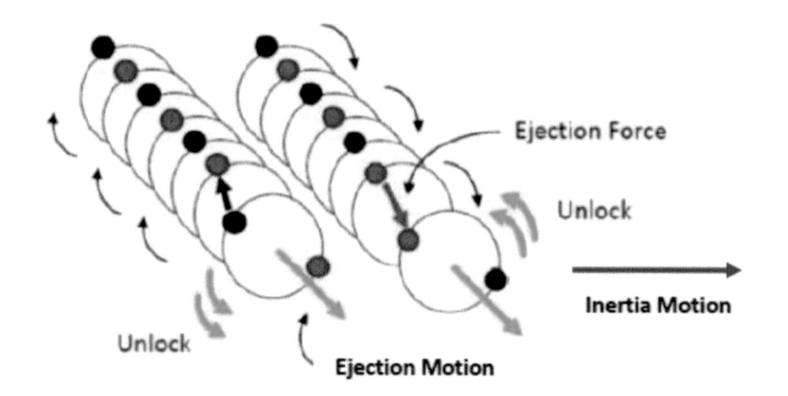

Fig. 19 A photon is formed in a two stage separation and ejection process by releasing Wu's Pair from its parent substance.

7.4. Absolute Light Speed

Because of the constant ejection force in the photon emission process, regardless to the frequency, a photon escaped from its parent object should always have a constant speed 3×10^8 m/s known as "Absolute Light Speed" in the ejection direction observed at the parent object (light source).

7.5. Photon Inertia Transformation

Photon emitted from an object carries the inertia of the parent object (light source). In other words, photon travels not only at an Absolute Light Speed (3×10^8 m/s) in the trajectory direction from the light source, but also with a speed and direction as that of the light source from the observer. This phenomenon is named "Photon Inertia Transformation" [29].

7.6. Wave Length, Momentum and Energy of Photon

Unlike Wu's Pairs having a momentum and energy as a portion of the substance, the momentum and energy of a free photon is generated during the two stage photon separation and ejection process from the light source. In other wors, the kinetic energy of a photon is equal to the energy difference between a photon emitted from an object and a Wu's Pair on the surface of the object.

Since energy is the amount of interaction applied on an object caused by the resistance of the action and the response of the action, therefore the kinetic energy (ΔE) of a photon can be represented by the multiplication of the momentum (P) and the change of velocity (ΔV) as follows:

$$\Delta E = P\Delta V$$

Because,

$\Delta E = E$

Also,

$\Delta V = C$

Therefore,

$$E = PC$$

Where E is the kinetic energy and P is the momentum of a photon and C is the light speed in space.

Because

$P = K_1\nu$

Therefore,

$E = K_1\nu C = (K_1C)\nu$

Given

$h = K_1C$

Therefore,

$$E = h\nu$$

Where E is the kinetic energy of a photon and h is Planck constant 6.626×10^{-34} m^2 kg/s.

Also,

$P = E/C = (h\nu/C) = h/(C/\nu)$

$C/\nu = \lambda$

Therefore,

$$P = h/\lambda$$

$$\lambda = h/P$$

Where P is the momentum and λ is the wavelength of a photon.

It is believed that during the photon separation process, the string energy associated with the string force between two Wu's Pairs on the surface of the substance is converted to the kinetic energy of the photon.

In summary, Table 1 shows the comparisons of energy and momentum between photons, Wu's Pairs, subatomic particles and string structures.

Table 1 Comparisons of energy and momentum between photons, Wu's Pairs and String Structures (Gravitons, etc.)

Photons	Wu's Pairs & String Structures (Gravitons, etc.)
$E = h\nu$	$E = \frac{1}{2} MV^2$
$P = h/\lambda$	$P = MV$
$E = PC$	$E = P^2/2M$

7.7. Mass of Photon (Wu's Pair)

Furthermore, the mass of a photon (or a Wu's Pair) can be calculated as follows [74]:

Because

$h = K_1C = m_{yy}KC$

Therefore,

$$m_{yy} = h/KC$$

Where m_{yy} is the mass of photon (Wu's Pair), h is Planck constant, K is whirlpool constant and C is Absolute Light Speed $3x10^8$ m/s.

7.8. Interference

The electromagnetic waves generated by a group of free photons can line up with each other to form a coherent light (Fig. 20). When two free photons come together, with their electromagnetic waves in phase, totally coinciding with each other, they are constructive. When they are 180 ° away from each other, they are destructive. This phenomenon is known as "Interference" [30]. When a beam of photons passes through a single slit, an interference pattern can be generated due to the interaction of the electromagnetic force between the photon (a free circulating Wu's Pair) and the edge of the slit. This phenomenon is known as "Single Slit Interference".

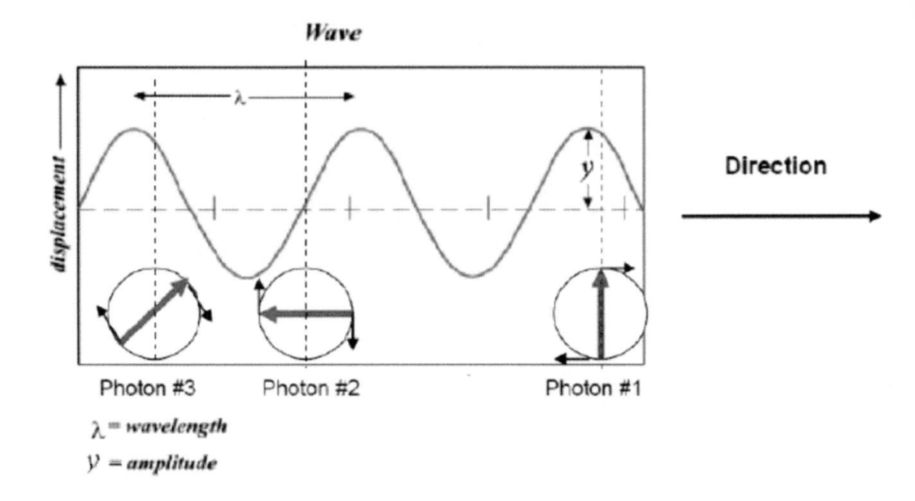

Fig. 20 Coherent photons and their electromagnetic waves.

7.9. Refraction and Reflection

Photon travels at different speeds in the matter of different densities. When crossing the boundary of two different matters, photon changes its direction in order to remain its coherency and frequency. This phenomenon is known as "Refraction" [31] [Annex 31]. However, on the other hand, photon can be bounced back from a flat surface like a particle. This phenomenon is known as "Reflection" [31].

7.10. Polarization

Photon travels like a flying disc in the plane of Yangton and Yington circulation. Photon can pass through a polarizer at a preferred orientation where both the plane of circulation and the direction of polarization are parallel. This phenomenon is known as "Polarization" [32].

7.11. Yangton and Yington Circulation

Fig. 21 is the schematic diagram of a Wu's Pair – a Yangton and Yington circulating Antimatter particle pair. The central acceleration (a_c) can be derived as follows:

$$a_c = dV/dt = (VdS/r)/dt = V(dS/dt)/r = V^2/r$$

Therefore,

$$F_c = ½\, m_{yy}\, a_c = ½\, m_{yy}\, V^2/r$$

Where m_{yy} is the mass of a single Wu's Pair.

Also, because of Coulomb's Law of Electrical Force,

$$F_{attraction} = k\, q_{yy}{}^2/(2r)^2$$

Wher k is Coulomb's Constant and q_{yy} is the charge of a Yangton particle or a Yington particle.

And

$$F_c = F_{attraction}$$

Therefore,

$$½\, m_{yy}\, V^2/r = k\, q_{yy}{}^2/(2r)^2$$

$$V^2r = ½\, k\, (q_{yy}{}^2/m_{yy})$$

Given

$$K = ½\, k\, (q_{yy}{}^2/m_{yy})$$

Therefore,

$$V^2r = K$$

Where K is named Wu Constant.

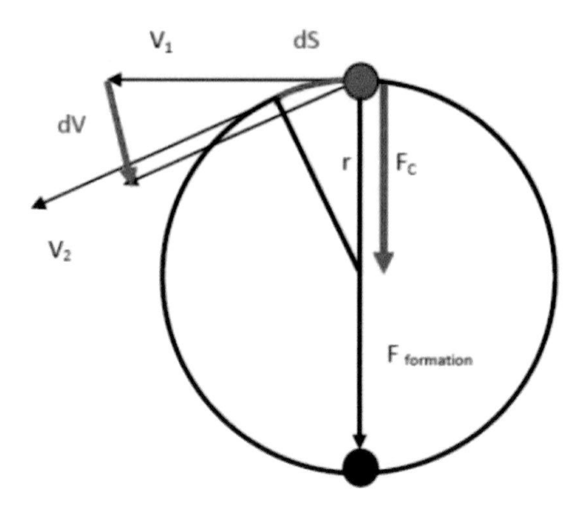

Fig. 21 Schematic diagram of a Wu's Pair.

7.12. Black Body Radiation

Fig. 22 shows the spectrums of Black Body Radiation [33] at different temperatures. It is obvious that at high temperatures, the peak of the spectrum corresponding to the amount of majority photons emitted from the Black Body, has higher energy (hν) but a smaller wavelength (λ). This can be explained well by Yangton and Yington circulating pairs as follows:

When temperature (T) increases (↑), then circulation energy ($E_{circulation}$) also increases (↑).

Because, $E_{circulation} = ½ MV^2$, therefore, velocity of circulation ($V_{circulation}$) also increases (↑).

Because, V^2r = constant, therefore, the radius of the circulation orbit (r) decreases (↓).

Because, circulating frequency equals to $V_{circulation}/2\pi r$, therefore, frequency (ν) increases (↑).

Also, energy of photon ($E_{Photon} = h\nu$) increases (↑).

Because $\nu\lambda = C$ = constant, therefore wavelength (λ) decreases (↓).

The energy distribution curve reflects a normal distribution of a group of quantized particles which proves the existence of photons.

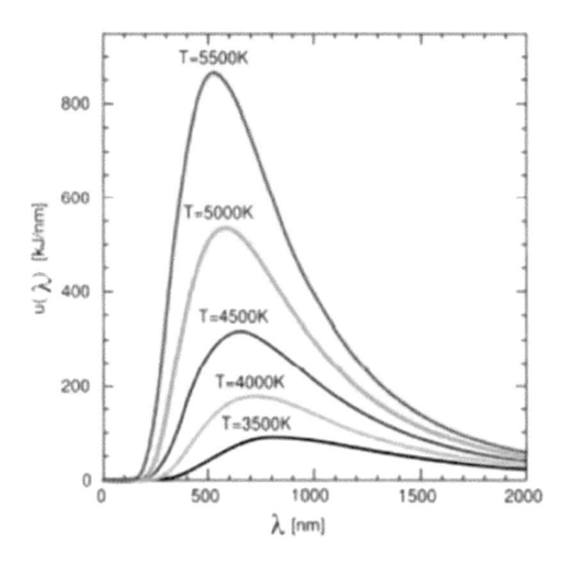

Fig. 22 Spectrums of Black Body Radiation at different temperatures.

Chapter Eight

Light Speeds

Is light Speed Truly a Constant?

8.1. Absolute Space System

In the universe, everything moves with respect to each others. There is no fixed reference. However, when a photon is emitted from a light source it generates a straight optical path from its light origin (not light source) into space. The light origin has an absolute fixed position in space that doesn't move with the light source, nor earth or anything else. Therefore, an Absolute Space System [34] can be defined by three fixed perpendicular axes (in reference to some far distance stars such as North Star) at the light origin.

8.2. Vision of Object

The "Vision of Object" is an object observed at a reference point (system) during a period of time. The object is correlated to the reference point (system) by distance and direction (or coordination). The reference point (system) has a fixed origin and three perpendicular axes.

8.3. Principle of Vision

The relative positions and directions between two objects maintain unchanged no matter of the reference points (systems). In other words, one object observed by the observer at a fixed reference point (system) on the other object maintains the same distance and direction (or coordination) no matter of the observation of the two objects at any other reference points (systems). This phenomenon is named "Principle of Vision" [35].

8.4. Theory of Vision

Based on Principle of Vision, a vision of object, in spite of observed directly at a reference point (system), can be constructed from the object and the reference point (system) in each time frame observed at the third reference point (system). A vision of object can be produced by superimposing the object observed in each time frame at the third reference point (system) while overlapping the reference point (system) observed in each time frame at the third reference point (system) perfectly on top of each other by keeping the same relative position and direction between the object and the reference point (system) as that observed at the third reference point (system). This is named "Theory of Vision" [35].

Two schematic diagrams are illustrated here to explain the construction process of vision of object from one reference point (system) to another reference point (system):

Fig. A shows the vision of an object observed at reference point O. Object t_1, Object t_2 and Object t_3 represent the positions and directions of the object; and Observer t_1, Observer t_2 and Observer t_3 represent the positions and directions of the observer, observed at the reference point O in the time frame t_1, t_2 and t_3 respectively. The curve from Object t_1 to Object t_2 and Object t_3 represents the vision of the object observed at reference point O during the time period from t_1 to t_3.

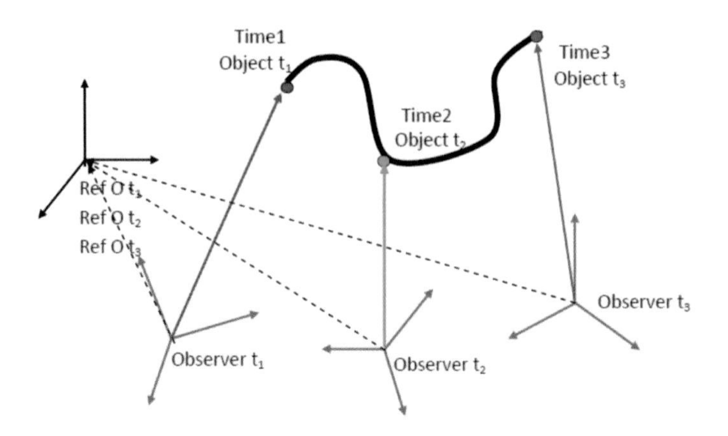

Fig. A Vision of an object observed at a reference point.

Fig. B shows the vision of object constructed at the final position of the observer from the vision of object observed at the reference point O. In which, Observer t_1, Observer t_2 and Observer t_3 and their coordination systems are completely matched and overlapped on top of Observer t_3. The same relative positions and directions of the Object t_1, Object t_2 and Object t_3 with respect to Observer t_1, Observer t_2 and Observer t_3 are maintained as that in Fig. A observed at reference point O. A curve from Object t_1 to Object t_2 and Object t_3 representing the vision of object observed by the observer during the time period from t_1 to t_3 can thus be constructed.

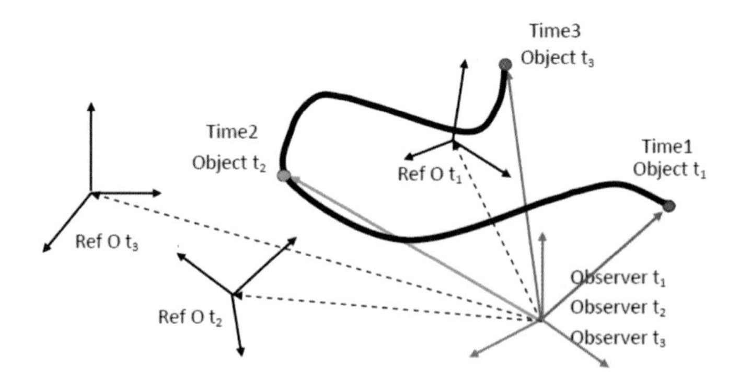

Fig. B Vision of an object observed at an observation point constructed from a reference point.

8.5. Vision of Light

Like vision of object, "Vision of Light" [34][35] is a photon observed at a reference point (system) during a period of time. The photon is correlated to the reference point (system) by distance and direction (or coordination).

Light speed is defined by the traveling distance of a photon divided by the traveling time measured at the reference point (system). This traveling distance of a photon is measured based on the Vision of Light observed by the observer during a period of time at the reference point (system). Therefore, light speed can be calculated as the Vision of Light divided by the traveling time of the photon measured at the reference point (system).

Similar to vision of object, in spite of observed directly by the observer at the reference point (system), Vision of Light can be constructed by superimposing the images of the photon and the reference point (system) observed at the light origin in the Absolute Space System. The reference point (system) of each time frame is overlapped on that of the final time frame. Also, the relative positions and directions between the photon and the reference point (system) are maintained as that observed at the light origin in the Absolute Space System.

Fig. 23 shows a schematic diagram of the Visions of Light of an emitted photon with respect to the observers at the light origin, ground and light source in Absolute Space System. Ground and light source drift apart from the light origin due to the motions of earth ($\mathbf{V_E}$) and the light source ($\mathbf{V_C}$). After a time interval Δt, assuming all motions are at constant speeds, the visions of light of those observers can be represented by the following straight lines: **AP**–the Vision of Light observed at light origin (black line), **BP**–the Vision of Light observed at ground (red line) and **CP**–the Vision of Light observed at the light source (green line) respectively. They all end at the final position point **P** of the emitted photon.

AP (Vision of Light observed at light origin) is the vector summation of **CP** (Vision of Light observed at the light source) and **AC** (moving path of the light source from light origin). Also, $\mathbf{C_O}$ (light speed observed at light origin) is the vector summation of $\mathbf{C_S}$ (light speed observed at the light source) and $\mathbf{V_C}$ (moving speed of the light source from light origin).

$$\mathbf{AP = CP + AC}$$

$$\mathbf{C_O = C_S + V_C}$$

Similarly, **BP** (Vision of Light observed at ground) is the vector summation of **CP** (Vision of Light observed at the light source) and **BC** (moving path of the light source from ground). Also, $\mathbf{C_E}$ (light speed observed at ground) is the vector summation of $\mathbf{C_S}$ (light speed observed at the light source) and $\mathbf{V_S}$ (moving speed of the light source from ground).

$$\mathbf{BP = CP + BC}$$

$$\mathbf{C_E = C_S + V_S}$$

Because of the constant repulsive force generated between photon and the adjacent Wu's Pairs on the surface of the light source in the photon emission process, a constant light speed $\mathbf{C_S}$ (Absolute Light Speed $3\text{x}10^8$ m/s) in the photon ejection direction can always be observed at the light source regardless of $\mathbf{V_C}$ and $\mathbf{V_S}$ which is in compliance with Photon Inertia Transformation Process [29].

When a photon observed at different observation points, the traveling times of the photon are the same ($\Delta t_E = \Delta t_S = \Delta t_O$), but the Visions of Light are different (**AP ≠ BP ≠ CP**). Since light speed is measured as the Vision of Light divided by the photon traveling time observed at the observation point, therefore the light speeds are different ($\mathbf{C_E \neq C_S \neq C_O}$) at different observation positions.

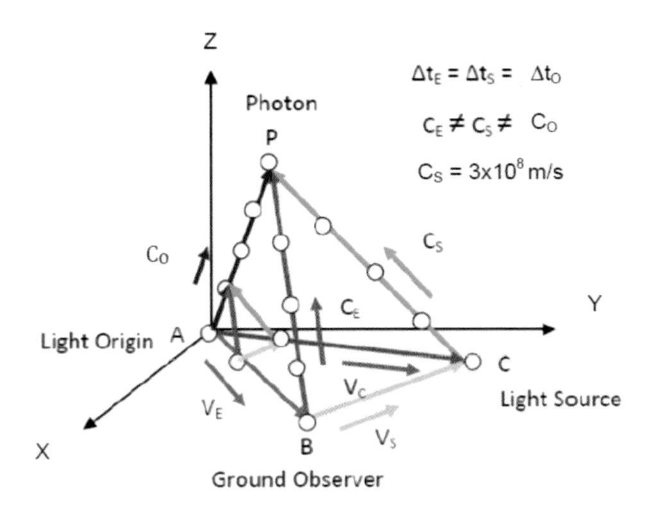

Fig. 23 Visions of Light of an emitted photon observed at the light origin (black line), ground (red line) and light source (green line) in Absolute Space System.

8.6. Equation of Light Speed

When a photon emitted from a light source, it travels under two influences, ejection motion and inertia motion. In other words, the light speed observed by the observer at any observation point (C) is a vector summation of the Absolute Light Speed 3 x 10^8 m/s, the moving speed of the photon away from the light source observed at the light source (C_S), and the "Inertia Light Speed", the moving speed of the light source away from the observer (or his inertia system) observed at the reference (observation) point (V_S). This theory is named "Equation of Light Speed" [35].

$$\mathbf{C} = \mathbf{C_S} + \mathbf{V_S}$$

Where $\mathbf{C_S}$ is the Absolute Light Speed and $\mathbf{V_S}$ is the Inertia Light Speed.

8.7. Michelson – Morley Experiment

In the 18th century, physicists believed that aether was the carrier of light (electromagnetic waves) in space. Michelson - Morley Experiment [36] was designed to prove the existence of aether by detecting the difference of light speeds through the optical interference caused by the motion of aether.

Fig. 24 illustrates the Michelson - Morley Experiment. AP is the Vision of Light observed at the light origin (red line) in Absolute Space System. BP is the Vision of Light observed at the light source (black line), which drifts away from its origin at speed V_E. When photons reach the semi-transparent mirror through Vision of Light BP observed at the light source, they split into two perpendicular light beams. These two beams are bounced back from the two end mirrors placed at equal distances from the center, then recombined at the semi-transparent mirror and finally received by the detector.

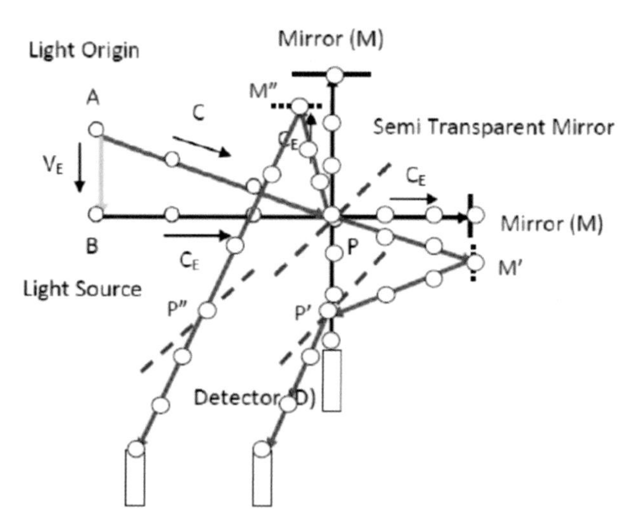

Fig. 24 Michelson – Morley Experiment with the Visions of Light observed at the light source (black line) and light origin (red line)

There are two ways to observe the experiment, from either the light origin or the light source. The result should be exactly the same. Here we try to observe the experiment from the light origin in Absolute Space System which is more complicated than that from the light source.

Because of the Photon Inertia Transformation, the light speed observed at the light source is always constant $C_E = 3 \times 10^8$ m/s and the light speed observed at light origin $\mathbf{C_o}$ is a vector summation of $\mathbf{C_E}$ and $\mathbf{V_E}$ (speed of light source away from light origin) .

$$\mathbf{C_o} = \mathbf{C_E} + \mathbf{V_E}$$

$C_O = (C_E^2 + V_E^2)^{1/2}$ (In case $\mathbf{C_E}$ is perpendicular to $\mathbf{V_E}$)

$C_O' = C_E - V_E$ (In case $\mathbf{C_E}$ and $\mathbf{V_E}$ are in opposite direction)

$C_O'' = C_E + V_E$ (In case $\mathbf{V_E}$ are in the same direction)

1. If aether doesn't exist –The time needed for photons traveling from P to M" and M" to P" is the same as that of traveling from P to M' and M' to P'.

Because,

$\Delta t = PM/C_E$

$\Delta t''_1 = PM''/(C_E - V_E) = (C_E\Delta t - V_E\Delta t)/(C_E - V_E) = \Delta t$

$\Delta t''_2 = M''P''/(C_E + V_E) = (C_E\Delta t + V_E\Delta t)/(C_E + V_E) = \Delta t$

$\Delta t'_1 = PM'/(C_E^2 + V_E^2)^{1/2} = ((C_E\Delta t)^2 + (V_E\Delta t)^2)^{1/2}/(C_E^2 + V_E^2)^{1/2} = \Delta t$

$\Delta t'_2 = M'P'/(C_E^2 + V_E^2)^{1/2} = ((C_E\Delta t)^2 + (V_E\Delta t)^2)^{1/2}/(C_E^2 + V_E^2)^{1/2} = \Delta t$

Therefore,

$\Delta t''_1 + \Delta t''_2 = \Delta t'_1 + \Delta t'_2 = 2\Delta t$

Where PM", M"P", PM' and M'P' are the lengths of light paths (visions of lights), and $\Delta t''_1$, $\Delta t''_2$, $\Delta t'_1$ and $\Delta t'_2$ are their respective photon traveling times.

2. If aether does exist – Michelson – Morley believed because of the flow of aether (V_A) the time needed for photons traveling from P to M" and M" to P" should be different from that traveling from P to M' and M' to P', so that an optical interference could be detected.

Because the traveling times are the same no matter the observation positions, they can be easily measured at light source instead of that at the light origin. Therefore,

$$\Delta t''_{A1} = PM''/(C_E - V_E + V_A) = PM/(C_E + V_A)$$

$$\Delta t''_{A2} = M''P''/(C_E + V_E - V_A) = MP/(C_E - V_A)$$

$$\Delta t'_{A1}=PM'/(C_E{}^2+(V_E - V_A)^2)^{1/2}=((C_E\Delta t)^2+(V_A\Delta t)^2)^{1/2}/(C_E{}^2+V_A{}^2)^{1/2} =\Delta t$$

$$\Delta t'_{A2}=M'P'/(C_E{}^2+(V_E-V_A)^2)^{1/2}=((C_E\Delta t)^2+(V_A\Delta t)^2)^{1/2}/(C_E{}^2+V_A{}^2)^{1/2} =\Delta t$$

Also,

$$\Delta t_{A1}'' + \Delta t_{A2}'' = PM\ (2C_E/(C_E{}^2-V_A{}^2) > 2PM/C_E = 2\Delta t$$

$$\Delta t'_{A1} + \Delta t'_{A2} = 2\Delta t$$

Therefore,

$$\Delta t_{A1}'' + \Delta t_{A2}'' > \Delta t'_{A1} + \Delta t'_{A2} = 2\Delta t$$

Where PM", M"P", PM' and M'P' are the lengths of light paths (visions of lights), and $\Delta t''_{A1}$, $\Delta t''_{A2}$, $\Delta t'_{A1}$ and $\Delta t'_{A2}$ are their respective photon traveling times under the influence of aether.

Since no optical interference was ever found in the experiment, Michelson and Morley concluded that aether doesn't exist in the universe. In fact, the Michelson – Morley Experiment has also proved that Photon Inertia Transformation exists and the light speed is always a constant observed in an inertia system.

8.8. Various Light Speeds in Space

8.8.1. Absolute Light Speed

During the photon emission process, the photon ejection force caused by the repulsion of the adjacent Wu's Pairs maintains the same regardless of the frequency (Fig. 19) (Fig. 25).

Therefore, photons travel in space at a constant Absolute Light Speed (3 x 10^8 m/sec) in their ejection directions, can always be observed at the light source and those stationary to the light source.

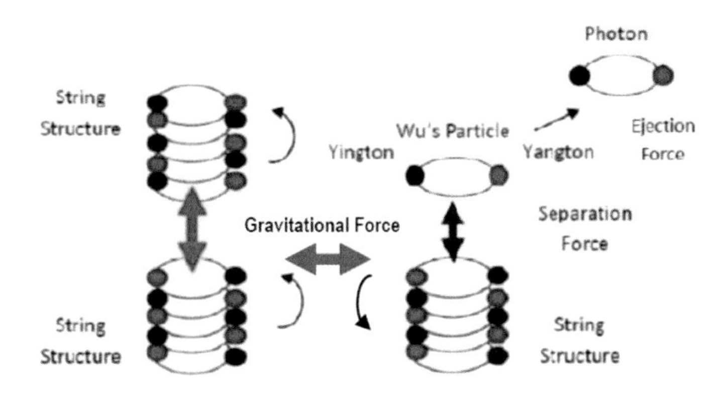

Fig. 25 Gravitational force between two gravitons (string structures) versus separation force between Wu's Pair and string structure

8.8.2. Light Speeds by Observations

The speed of light is calculated by the Vision of Light divided by the traveling time of a photon. Since different Visions of Light of a photon can be observed by observers at different moving speeds and directions, it is obvious that different light speeds in space can be observed by moving observers other than those at the light source. As shown in Fig. 23, in addition to the Absolute Light

Speed C_S (3 x 10^8 m/s) observed by at the light source, light speeds C_E and C_O can also be observed at ground and light origin. This is different from that of Einstein's Special Relativity, which claims that light speed in space is always constant, no matter the directions and speeds of the light sources and observers.

Furthermore, if an observer moves at a speed as fast as the Absolute Light Speed, in a parallel direction to the light beam, then the light speed observed by the moving observer can be as small as zero. In other words, the photon is frozen or idles with respect to the observer [35].

Because,

$$\mathbf{C = C_S + V_S}$$

$$\mathbf{V_S = - C_S}$$

Therefore,

$$\mathbf{C = 0}$$

In addition, Einstein claimed that if he was running with a photon at light speed, he could still see the photon moving away from him at the light speed. It is impossible, unless he was running with the light source at a light speed away from the light origin [35].

Because,

$$\mathbf{C = C_S + V_S}$$

$$\mathbf{V_S = 0}$$

Therefore,

$$\mathbf{C = C_S}$$

8.8.3. Light Speeds in Inertia System

As a photon traveling in space (Fig. 26), the same Visions of Light (red line) can be observed at different positions in an Inertia System. Therefore, the light speeds observed in an inertia system should always be constant, no matter where the observers are. However this constant speed can be different from the Absolute Light Speed (3 x 10^8 m/s) if it is observed in a different Inertia System other than that of the light source.

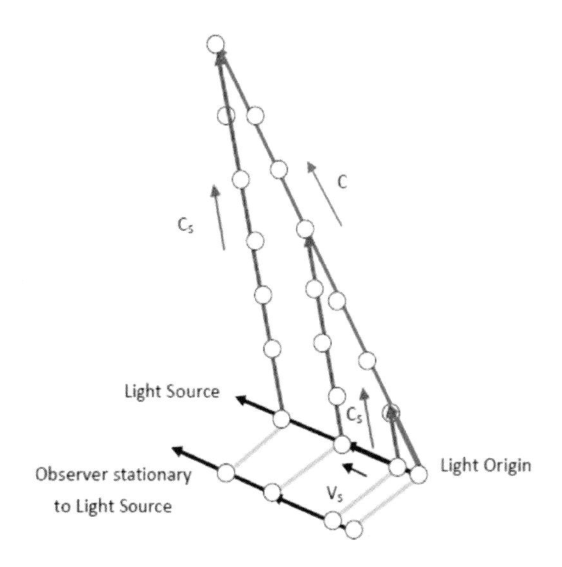

Fig. 26 Vision of Light (blue line) observed at light origin and the Vision of Light (red line) observed at the light source and those stationary to the light source.

8.8.4. Light Speeds on Earth

C_E (light speed observed at ground) is the vector summation of C_S (light speed observed at the light source) and V_S (moving speed of the light source from ground).

$$\mathbf{C_E = C_S + V_S}$$

However, in reality, V_S the moving speed of the light source from ground is extremely small compared to C_S the Absolute Light Speed (3x 10^8 m/s). Therefore,

$$C_S \gg V_S$$

$$C_E = C_S + V_S \approx C_S$$

The light speed observed by the ground observer C_E is very close to the Absolute Light Speed C_S.

8.8.5. Limit of Light Speed

Because Wu's Pairs are the finest building blocks of all matter, when a Wu's Pair separates from the surface of a substance (string structures) to form a free photon, it can be accelerated by the repulsive force between the two Yangton particles (one from the emitting photon and the other one from the adjacent Wu's Pair) also that between the two Yington particles (one from the emitting photon and the other one from the adjacent Wu's Pair) on the surface of the parent substance, to reach an extremely high speed (3×10^8 m/s). Therefore, it is suggested that the Absolute Light Speed 3×10^8 m/s is the highest speed any object can move in the universe. However, in theory, there should be no limit but subject to the driving force.

8.9. Gravitational Lensing

When a light beam passes through a massive star (or dark matter), its path is curved due to the gravitational field. This phenomenon is known as "Deflection of Light" [78][80][Annex 31]. According to general relativity, it is resulted from the the distortion of spacetime. However, based on Yangton and Yington Theory, it is caused by the expansion of photon (large wavelength) and the reduction of light speed. Just like the light beam passing through a transparent material, the light speed is reduced and the direction is curved in order to maintain the same frequency and coherency. The massive star works like a telescope, the deflected light beams from a star (light origin) behind the massive star can be focused into several images such that a clear picture of the star (light origin) can be observed and the distance of the star (light origin) and the mass of the massive star can be calculated. This is called "Gravitational Lensing" [37].

8.10. Doppler Effect

The frequency of a wave-like signal such as sound or light depends on the movement of the sender and the receiver. This phenomenon is known as the "Doppler Effect" [38]. When the source of light is moving toward the observer, each successive photon is emitted from a position closer to the previous one [34]. In other words, assuming that light emission is a Non-Inertia Transformation, the wavelength between two subsequent photons is smaller, which causes an increase in the frequency or shift to the blue end of the spectrum. This is commonly known as "Blue Shift". Conversely, if the source of light is moving away from the observer, each photon is emitted from a position farther from the previous photon, resulting in long wavelengths between the two subsequent photons. This causes a reduction in the frequency or shift towards the red end of the spectrum, which is known as "Redshift" (Fig. 27).

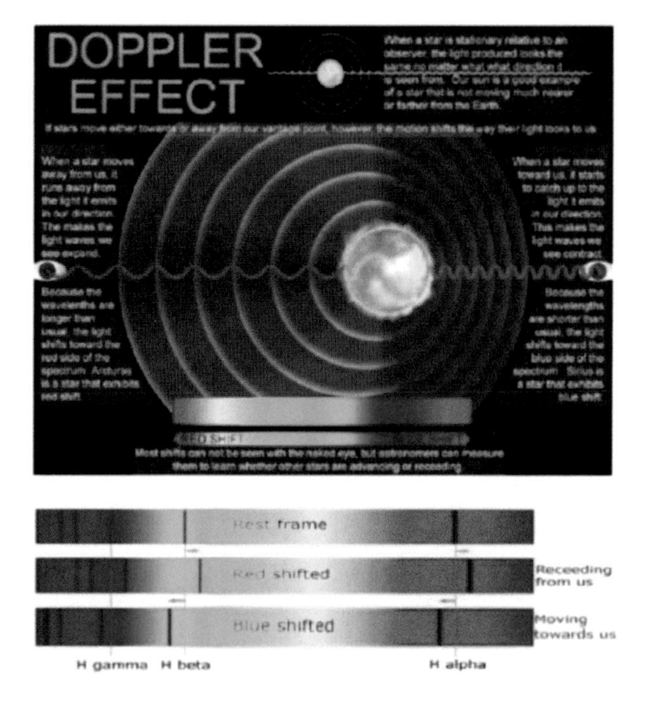

Fig. 27 Doppler Effect and Blueshift & Redshift phenomena.

Like most scientists, I first thought that the Redshift and the Doppler Effect could only exist in the Non-Inertia Transformation process [34] as that in sound propagation. I therefore believe that a photon emitted from a light source travels in space at an Absolute Light Speed 3 x 10^8 m/s can be observed at its origin in Absolute Space System, without any influence from its light source, as noted, in my previous publication [34]. This, however, is in conflict with my logical thinking. It is hard to believe that a ball thrown out of the window of a train will not follow the train. The concept that photon emission is a Non-Inertia Transformation bothered me for a long time until I developed the Acceleration Doppler Effect based on the Photon Inertia Transformation process to solve the problem.

8.11. Acceleration Doppler Effect

The Doppler Effect can be proven easily in the Non-Inertia Transformation process with the signal source traveling at a constant speed [34] either towards or away from the observer such as that of sound propagation. However, the photon emission from the light source is an Inertia Transformation process (Fig. 23) [29]. Both Redshift and Blue shift occurred only when the wavelength of light changes with the acceleration of the light source. I call this phenomenon "Acceleration Doppler Effect" [29].

For a star far away from earth, the ground observer is considered in stationary to the light origins of all photons that emitted from the light source (star). Therefore the Vision of Light [35] of each photon observed by the ground observer is the same as that observed at the light origin of the photon in the Absolute Space System.

The light source (star) can either move toward or away from the observer on earth. Assuming it takes time t for a photon traveling from light source (star) to earth. V_o is the speed of the light source (star) at its beginning, V_t is the speed of the light source (star) at time t and a is the constant acceleration of the light source (star) in time t. S is the distance of the light source (star) traveling from the light origin in time t. P is the distance of the photon traveling from the light origin to earth in time t, $V_o t$ is the distance of the photon dragged by the light source (star) in time t and D is the distance between the light source (star) and the photon when it reaches earth at time t. Also λ_1 is the wavelength, ν_1 is the frequency and C_1 is the light speed of the photon from light origin observed on light origin and earth. Zeroshift, Blueshift and Redshift phenomena resulted from the Acceleration Doppler Effect (Fig. 28) can thus be derived as follows:

OS = **S** = Distance from light origin to light source = Motion of light source away from light origin.

SP = **D** = Distance from light source to photon = Vision of light observed from light source.

OP = **P** = Distance from light origin to photon = Vision of light observed from light origin and ground.

$$\mathbf{OP = OS + SP}$$

$$\mathbf{P = S + D}$$

$$\mathbf{D = P - S}$$

Also,

$$\mathbf{OP = P = Ct + V_0}t$$

Where **C** is the Absolute Light Speed, $\mathbf{V_0}$ is the initial moving speed of light source from light origin and t is time.

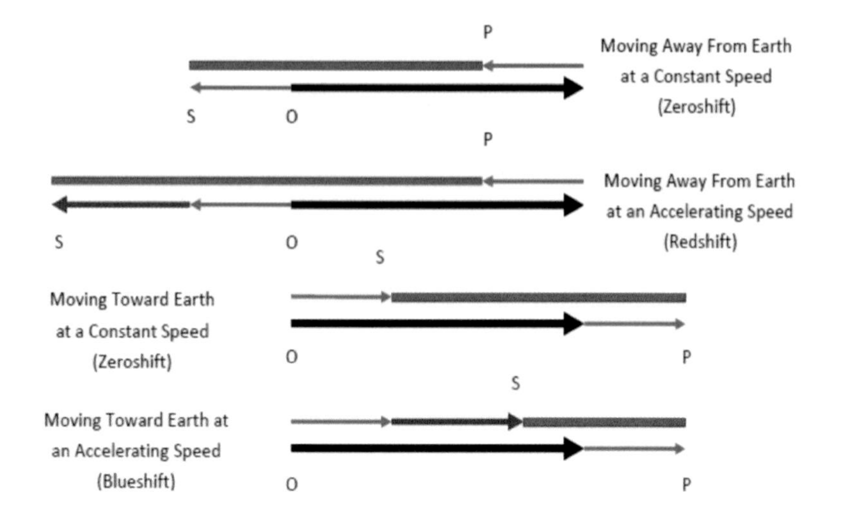

Fig. 28 Zeroshift, Redshift and Blueshift caused by Acceleration Doppler Effect.

A. Zeroshift

When the light source (star) either moves toward or away from the observer on earth at a constant speed ($V_o = V_t$ and $a = 0$), Zeroshift can be observed.

1. In case the light source (star) moves away from the observer on earth,

$S = - V_o t$

$P = Ct - V_o t$

$D = P - S = Ct$

Therefore,

$\lambda_1 = D/\nu t = Ct/\nu t = C/\nu = \lambda$

$C_1 = P/t = (Ct - V_o t)/t = C - V_o < C$

$\nu_1 = C_1/\lambda_1 = (C - V_o)/\lambda < \nu$

When the light source (star) moves away from the observer on earth at a constant speed, the wavelength maintains unchanged, but both frequency and light speed become smaller. Zeroshift can be observed.

2. In case the light source (star) moves toward the observer on earth,

$S = V_0t$

$P = Ct + V_0t$

$D = P - S = Ct$

Therefore,

$\lambda_1 = D/\nu t = Ct/\nu t = C/\nu = \lambda$

$C_1 = P/t = (Ct + V_0t)/t = C + V_0 > C$

$\nu_1 = C_1/\lambda_1 = (C + V_0)/\lambda > \nu$

When the light source (star) moves toward the observer on earth at a constant speed, the wavelength maintains unchanged, but both frequency and light speed become bigger. Zeroshift can be observed.

B. Blueshift

In case the light source (star) moving toward the observer on earth at a constant acceleration speed,

$S = V_0t + ½ at^2$

$P = Ct + V_0t$

$D = P - S = Ct - ½ at^2$

Therefore,

$\lambda_1 = D/\nu t = (Ct - \frac{1}{2} at^2)/\nu t = (C - \frac{1}{2} at)/\nu < \lambda$

$C_1 = P/t = (Ct + V_o t)/t = C + V_o > C$

$\nu_1 = C_1/\lambda_1 = (C + V_o)/((C - \frac{1}{2} at)/\nu) > \nu$

When the light source (star) moves toward the observer on earth at a constant acceleration speed, the wavelength becomes smaller, both the frequency and light speed become bigger, and thus Blueshift can be observed.

C. Redshift

In case the light source (star) moving away from the observer on earth at a constant acceleration speed,

$S = -(V_o t + \frac{1}{2} at^2)$

$P = Ct - V_o t$

$D = P - S = Ct + \frac{1}{2} at^2$

Therefore,

$\lambda_1 = D/\nu t = (Ct + \frac{1}{2} at^2)/\nu t = (C + \frac{1}{2} at)/\nu > \lambda$

$C_1 = P/t = (Ct - V_o t)/t = C - V_o < C$

$\nu_1 = C_1/\lambda_1 = (C - V_o)/((C + \frac{1}{2} at)/\nu) < \nu$

When the light source (star) moves away from the observer on earth at constant acceleration speed, the wavelength becomes bigger, both the frequency and light speed become smaller, and thus Redshift can be observed.

8.12. Redshifts

According to the Acceleration Doppler Effect, Doppler Redshift occurs whenever a light source (star) moves away from the observer on earth at an acceleration speed. There are another two Redshifts in the universe. Gravitational Redshift [40] [41] can be observed for a photon emitted from a star of massive gravitational field. Cosmological Redshift [41] [42], on the other hand, can be observed for a photon emitted from a star of several billion light years away.

8.13. Event Horizon

When a light source accelerating towards the center of a black hole, because of the Photon Inertia Transformation, the photon emitted from the light source compiles two competing opposite speeds: (1) outward Absolute Light Speed ($\mathbf{C_S}$) and (2) inward Inertia Light Speed ($\mathbf{V_S}$).

According to Equation of Light Speed,

$$\mathbf{C} = \mathbf{C_S} + \mathbf{V_S}$$

At Event Horizon,

$|\mathbf{C_S}| = |\mathbf{V_S}|$, therefore $\mathbf{C} = \mathbf{0}$.

Inside Event Horizon,

$|\mathbf{C_S}| < |\mathbf{V_S}|$, therefore $\mathbf{C}$ follows $\mathbf{V_S}$ and goes inwards.

Outside Event Horizon,

$|\mathbf{C_S}| > |\mathbf{V_S}|$, therefore $\mathbf{C}$ follows $\mathbf{C_S}$ and goes outwards.

Where $\mathbf{C}$ is the light speed observed by observer at the light origin, $\mathbf{C_S}$ is the Absolute Light Speed observed at the light source and $\mathbf{V_S}$ is the speed of light source observed at the light origin (Inertia Light Speed).

As a result, at the Event Horizon [76], the net speed of the photon is zero and the photon is in idle. Outside the Event Horizon (Ergosphere), the Absolute Light Speed is bigger than the Inertia Light Speed, the photon can move outwards and escape from the black hole. However, inside the Event Horizon, the Absolute light Speed is smaller than the Inertia Light Speed, the photon moves inwards and can never escape from the black hole (Fig. J). Therefore, the existence of a "Black Hole" can be predicted.

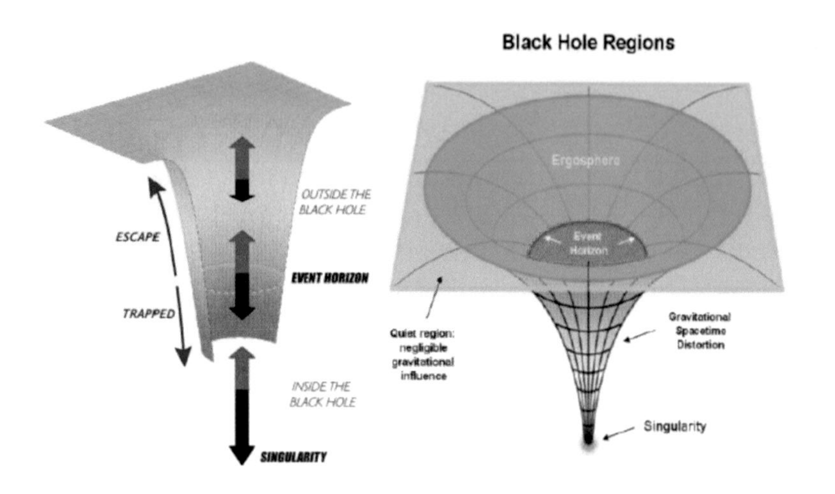

Fig. J The Black Hole Regions: (1) Ergosphere (C > V) (2) Event Horizon (C = V) and (3) Inside the Black Hole (C < V). C is light speed and V is the light source speed.

8.14. Length Contraction

The length of an object measured by an observer, along the direction of motion, is shorter than the length measured by another observer at rest with the object (Fig. 29). This phenomenon is known as "Length Contraction" [43]. It is caused by the difference of Visions of Light between a moving observer and a ground observer, rather than the Velocity Time Dilation [44] that is derived from Einstein's Special Relativity [39] which actually never exists.

Assume it takes time Δt for a photon to travel from the beginning to the end of an object with length L_E measured by a ground observer at the light source. The Visions of Light of a moving observer and a ground observer are L_S (red line) and L_E (black line), and the light speeds measured by the moving observer and ground observer are C_S and C_E respectively. The Length Contraction can be calculated as follows:

According to the Equation of Light Speed, the velocity of the photon observed by the observer $\mathbf{C_S}$ is the vector summation of the velocity of the photon observed at the light source $\mathbf{C_E}$ (here the light source is stationary on the ground), and the velocity of the light source moves away from the observer **V** (the negative velocity of the observer moves away from the light source –V). Therefore,

$\mathbf{C_S = C_E + V}$

$C_S = C_E + (-V)$

$L_E = C_E\,\Delta t$ & $L_S = C_S\,\Delta t$

$L_S = L_E + (-V)\,\Delta t$

$L_S/L_E = (L_E - V\Delta t)/L_E$

$L_S/L_E = (C_E - V)\,\Delta t / C_E \Delta t = (C_E - V)/C_E$

$L_S = (1 - V/C_E)\,L_E$

In case $0 < V < C_E$ (moving toward the end of object)

Then $1 > (1 - V/C_E) > 0$

And

$$L_S < L_E$$

In case $V < 0$ (moving away from the end of the object)

Then $1 - V/C_E > 1$

And

$$L_S > L_E$$

Depending on the moving speed and direction of the observer, the length of an object, which is equal to the Vision of Light observed by the moving observer, can either shrink or expand. In other words, both Length Contraction and Length Expansion can happen subject to the moving speeds and directions of the observers with respect to the object (Fig. 29).

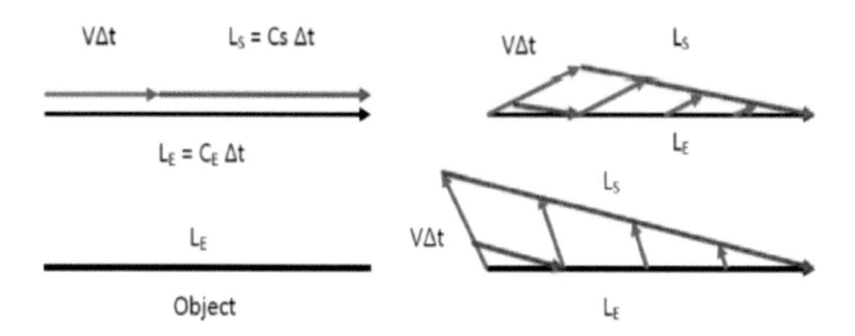

Fig. 29 The Visions of Light (red lines) observed at different moving speeds and directions (blue lines), with respect to the Vision of Light (black line) observed at ground.

Chapter Nine

Time and Space

What Is Time?

What Is Space?

Does Time Change with Speed?

Does Time Change with Gravity?

9.1. Definition of Time

"Time" is the duration of an event. It is a "Nature Quantity" and has an absolute value. Time can be measured by a "Unit Time" and represented by the "Amount of Unit Time" multiplied by the "Unit Time" which is known as "Measured Quantity" [Annex 28]. Unit Time is also a nature quantity which is the period of a specific repeating process such as the circulation period of Wu's Pairs (Wu's Unit Time) and electronic transition in an atomic clock. The time duration of an event doesn't change, but with different measurements, the amount of unit time could be different subject to the duration of the Unit Time. For the corresponding identical objects and events, the Amount of Corresponding Identical Unit Time keeps the same, but the Corresponding Identical Unit Time could be different subject to the gravitational field and aging of the universe.

9.2. Definition of Length

Similarly, "Length" is the size of an object. It is a nature quantity and has an absolute value. Length can be measured by a "Unit Length", and represented by the "Amount of Unit Length" multiplied by the "Unit Length" as the measured quantity [Annex 28]. Unit Length is also a nature quantity which is the length of a specific object such as the diameter of Wu's Pairs (Wu's Unit Length) and human's foot. The length of an object doesn't change, but with different measurements, the amount of Unit Length could be different subject to the size of the Unit Length. For Corresponding Identical Objects and Events, the Amount of Corresponding Identical Unit Length keeps the same, but the Corresponding Identical Unit Length could be different subject to the gravitational field and aging of the universe.

9.3. The Basic Units of Wu's Pairs

The measurement of a physical quantity such as mass, time and length contain two components: Unit and the Amount of Unit.

Since Wu's Pairs are the building blocks of all matter, a Wu's Pair (Wu's Unit Mass m_{yy}) can be used as the basic unit mass. Also, the circulation period of Wu's Pair (Wu's Unit Time t_{yy}) and the diameter of Wu's Pair (Wu's Unit Length l_{yy}) can be used as the basic unit time and basic unit length for the measurements of the objects and events at the same location with the same gravitational field and aging of the universe [41].

Because of the Conservation of Mass, Wu's Unit Mass m_{yy}, the mass of a single Wu's Pair, stays unchanged at all time. However, according to Wu's Spacetime Theory [41] that Wu's Unit Time depends on Wu's Unit Length ($t_{yy} = \gamma\, l_{yy}^{3/2}$), also basing on Wu's Spacetime Shrinkage Theory [67] that Wu's Unit Length increases with the gravitational field and decreases with the aging of the universe, Wu's Unit Time and Wu's Unit Length could be different from one location to the other location subject to the gravitational field and aging of the universe at the reference point.

9.4. Corresponding Identical Object and Event

When an object or event takes place or moves to a different location under an equilibrium condition, it keeps all of its properties in a corresponding state while maintaining still the same mass. In other words, it keeps all of its properties with the same "Amounts of Quantities", no matter the changes of "Unit Quantities" caused by the gravitational field and aging of the universe. This object is called "Corresponding Identical Object" and event is called "Corresponding Identical Event".

Corresponding identical object likes a stretched rope of rubber bands. Each rubber band has a unit length. The total amount of rubber bands doesn't change, but the length of each rubber band (corresponding identical unit length) and the total length of the rope could be different subject to the stretching force. Corresponding identical object also likes the giant in "Jack and the Beanstalk", and the dwarf in "Snow White", they have the same features as that of a normal man except in different sizes.

Corresponding identical event on the other hand likes a motion pictures, where each picture runs by a unit time, the total amount of pictures doesn't change, but the duration of each picture (corresponding identical unit time) and the total playing time could be different subject to the moving speed. Corresponding identical event also likes the Mickey Mouse cartoon pictures, the entire show can be completed by different time durations subject to the rolling speed of the pictures.

9.5. Principle of Correspondence

For a corresponding identical object or event, its physical property can be measured by the corresponding identical unit of the property. The amount of corresponding identical unit of the property always maintains the same no matter of the gravitational field and aging of the universe. This phenomenon is known as "Principle of Correspondence" [65].

Principle of Correspondence is only true based on the following two assumptions:

1. Wu's Pairs are the building blocks of all the matter in the universe (a single Wu's Pair has one fixed Wu's Unit Mass).

2. Wu's Pair always exists and can't be separated and destroyed by any means (conservation of mass).

Because of the Principle of Correspondence, all physical laws maintain unchanged in an inertia system (constant speed) of constant gravitational field and ageing of the universe. This is called "Law of Inertia".

9.6. Principle of Time

Based on the principle of correspondence, corresponding identical events measured by the corresponding identical unit time should have a constant amount of corresponding identical unit time, no matter where the event takes place and how the corresponding identical unit time is different from one location to the other. This theory is named "Principle of Time". For example, a 3000 cycles pendulum swing event on Saturn takes the same amount of cycles but more slowly than that on earth because the pendulum swing on Saturn is slower with longer period (Saturn second) than that on earth (Earth second), due to Saturn's large gravity.

9.7. Principle of Length

Similarly, a corresponding identical object measured by the corresponding identical length should have a constant amount of corresponding identical unit length, no matter where the object is and how the corresponding identical unit length is different from one location to the other. This theory is named "Principle of Length". For example, a man on Saturn can have the same six foot height but actually be taller than his twin on earth, because one foot on Saturn (Saturn foot) is longer than that on earth (Earth foot) due to Saturn's large gravity.

9.8. Wu's Time and Normal Time

9.8.1. Wu's Time

Since Wu's Pairs are proposed as the building blocks of all matter, the duration of an event called "Wu's Time" (T_w) can be measured by Wu's Unit Time (t_{yy}), the circulation period of a Wu's Pair at the same location, and to be represented by the Amount of Wu's Unit Time (a) multiplied by Wu's Unit Time [41].

$$T_w = a\ t_{yy}$$

According to the Principle of Time, for a corresponding identical event measured by the Corresponding Identical Wu's Unit Time, the Amount of Corresponding Identical Wu's Unit Time "a" should be a constant, no matter the gravitational field and aging of the universe.

9.8.2. Normal Time

The duration of an event called "Normal Time" (T_n) can be measured by a specific "Normal Unit Time" (t_s) such as "second" at the same location, and to be represented by the Amount of Normal Unit Time (t) multiplied by the Normal Unit Time [41].

$$T_n = t\ t_s$$

According to the Principle of Time, for a corresponding identical event measured by the Corresponding Identical Normal Unit Time, the Amount of Corresponding Identical Normal Unit Time "t" should be a constant, no matter the gravitational field and aging of the universe.

For example, a Cesium oscillator has an oscillation period 1/9,192,631,770 seconds (Earth second). The corresponding identical oscillator on Mars will also have an oscillation period of 1/9,192,631,770 seconds (Mars second). However the second on Mars is a Mars second instead of an Earth second.

9.9. Wu's Length and Normal Length

9.9.1. Wu's Length

Since Wu's Pairs are proposed as the building blocks of all matter, the length of an object called "Wu's Length" (L_w) can be measured by Wu's Unit Length (l_{yy}), the size of the circulation orbit of a Wu's Pair at the same location, and to be represented by the Amount of Wu's Unit Length (e) multiplied by Wu's Unit Length [41].

$$L_w = e l_{yy}$$

According to the Principle of Length, for a corresponding identical object measured by the Corresponding Identical Wu's Unit Length, the Amount of Corresponding Identical Wu's Unit Length "e" should be a constant, no matter the gravitational field and aging of the universe.

9.9.2. Normal Length

The length of an object called "Normal Length" (L_n) can be measured by a specific "Normal Unit Length" (l_s) such as "meter" at the same location, and to be represented by the Amount of Normal Unit Length (l) multiplied by the Normal Unit Length [41].

$$L_n = l\, l_s$$

According to the Principle of Length, for a corresponding identical object measured by the Corresponding Identical Normal Unit Length, the Amount of Corresponding Identical Normal Unit Length "l" should be a constant, no matter the gravitational field and aging of the universe.

For example, a one foot ruler has a length on earth of 30.48 cm (Earth centimeter). The corresponding identical ruler on Mars will also have a length of 30.48 cm (Mars centimeter). However, the centimeter on Mars is a Mars centimeter not that of an Earth centimeter.

9.10. Velocity, Normal Velocity and Wu's Velocity

Velocity is a quantity related to the change of the position of an object with respect to time. Velocity is defined by the distribution of an infinitesimal traveling length over an infinitesimal traveling time [Annex 29].

$$V = \Delta l/\Delta t$$

Where V is velocity, Δl is infinitesimal traveling length and Δt is infinitesimal traveling time.

9.10.1. Normal Velocity

The velocity of a motion called "Normal Velocity" (V_n) can be expressed by the Amount of Normal Unit Velocity (v) multiplied by the Normal Unit Velocity (l_s/t_s) such as "cm/s".

$$V_n = v\ (l_s/t_s)$$

Because,

$T_n = x\ t_s$ and $T_n' = x'\ t_s$

$L_n = y\ l_s$ and $L_n' = y'\ l_s$

$V_n = (L_n'-L_n) / (T_n'-T_n) = ((y'-y) / (x'-x))\ l_s/t_s$

Given

$v = (y'-y) / (x'-x)$

Therefore,

$$V_n = v\ (l_s/t_s)$$

Where V_n is Normal Velocity, v is Amount of Normal Unit Velocity and l_s/t_s is Normal Unit Velocity.

According to the Principle of Correspondence, for a corresponding identical motion measured by the Corresponding Identical Normal Unit Velocity, the Amount of Corresponding

Identical Normal Unit Velocity "v" should be a constant, no matter the gravitational field and aging of the universe.

9.10.2. Wu's Velocity

Silimarly, the velocity of a motion called "Wu's Velocity" (V_w) can be expressed by the Amount of Wu's Unit Velocity (w) multiplied by Wu's Unit Velocity (l_{yy}/ t_{yy}).

$$V_w = w\ (l_{yy}/t_{yy})$$

Where V_w is Wu's Velocity, w is Amount of Wu's Unit Velocity and l_{yy}/t_{yy} is Wu's Unit Velocity.

According to the Principle of Correspondence, for a corresponding identical motion measured by the Corresponding Identical Wu's Unit Velocity, the Amount of Corresponding Identical Wu's Unit Velocity "w" should be a constant, no matter the gravitational field and aging of the universe.

9.11. Acceleration, Normal Acceleration and Wu's Acceleration

Acceleration is a quantity related to the change of the velocity of an object. Acceleration is defined by the distribution of an infinitesimal velocity over an infinitesimal traveling time [Annex 29].

$$A = \Delta V/\Delta t$$

Where A is acceleration, ΔV is infinitesimal velocity and Δt is infinitesimal traveling time.

9.11.1. Normal Acceleration

The acceleration of a motion called "Normal Acceleration" (A_n) can be expressed by the Amount of Normal Unit Acceleration (a) multiplied by the Normal Unit Acceleration (l_s/ t_s^2) such as "cm/s^2".

$$A_n = a\ (l_s/t_s^2)$$

Because,

$T_n = x\, t_s$ and $T_n' = x'\, t_s$ and $T_n'' = x''\, t_s$

$L_n = y\, l_s$ and $L_n' = y'\, l_s$ and $L_n'' = y''\, l_s$

$V_n = (L_n' - L_n) / (T_n' - T_n) = ((y'-y) / (x'-x))\, l_s/t_s$

$V_n' = (L_n'' - L_n') / (T_n'' - T_n') = ((y''-y') / (x''-x'))\, l_s/t_s$

$A_n = (V_n' - V_n)/(T_n'' - T_n')$

$= [(y''-y') / (x''-x') - (y'-y) / (x'-x)]\, (l_s/t_s)/(x'' - x')t_s$

$= \{[(y''-y') / (x''-x') - (y'-y) / (x'-x)] /(x'' - x')\}\, (l_s/t_s)/t_s$

Given

$a = [(y''-y') / (x''-x') - (y'-y) / (x'-x)] / (x'' - x')$

Therefore,

$$A_n = a\, (l_s/t_s^2)$$

Where A_n is Normal Acceleration, "a" is Amount of Normal Unit Acceleration and l_s/t_s^2 is Normal Unit Acceleration.

According to the Principle of Correspondence, for a corresponding identical motion measured by the Corresponding Identical Normal Unit Acceleration, the Amount of Corresponding Identical Normal Unit Acceleration "a" should be a constant, no matter the gravitational field and aging of the universe.

9.11.2. Wu's Acceleration

Silimarly, the acceleration of a motion called "Wu's Acceleration" (A_w) can be expressed by the Amount of Wu's Unit Acceleration (b) multiplied by Wu's Unit Acceleration (l_{yy}/ t_{yy}^2) [41].

$$A_w = b\ (l_{yy}/t_{yy}^2)$$

Where A_w is Wu's Acceleration, b is Amount of Wu's Unit Acceleration and l_{yy}/t_{yy}^2 is Wu's Unit Acceleration.

According to the Principle of Correspondence, for a corresponding identical motion measured by the Corresponding Identical Wu's Unit Acceleration, the Amount of Corresponding Identical Wu's Unit Acceleration "b" should be a constant, no matter the gravitational field and aging of the universe.

In conclusion, Wu's Time (T_w), Wu's Length (L_w), Wu's Velocity (V_w) and Wu's Acceleration (A_w), like Normal Time (T_n), Normal Length (L_n), Normal Velocity (V_n) and Normal Acceleration (A_n) are all measured quantities of an object or event. Despite of the measurement methods, for the same object and event, all the measured quantities (Amount of Unit x Unit) are the same as that of the nature quantities (T, L, V, A). However, for the measured quantities, the Amounts of Unit Quantities could be different subject to the reference system. For example, the Amount of Wu's Unit Length, Wu's Unit Time, Wu's Unit Velocity and Wu's Unit Acceleration of an object or event measured by Wu's Unit Length (l_{yy}), Wu's Unit Time (t_{yy}), Wu's Unit Velocity (l_{yy}/t_{yy}) and Wu's Unit Acceleration (l_{yy}/t_{yy}^2) at a reference point on other planet are different from that measured by Wu's Unit Length (l_{yy0}), Wu's Unit Time (t_{yy0}), Wu's Unit Velocity (l_{yy0}/t_{yy0}) and Wu's Unit Acceleration (l_{yy0}/t_{yy0}^2) referenced on earth.

9.12. Einstein's Special Relativity

In Einstein's Special Relativity [39], it is postulated that the light speed in space is always a constant, no matter the motions and positions of the light sources and observers. As a consequence, time on a moving object runs slower than that of one stationary to the observer known as "Velocity Time Dilation" [44].

According to Yangton and Yington Theory, because the light speed in space is not always a constant, it changes according to the relative speeds and directions between the light source and the observers, Einstein's Special Relativity is false, as is the Velocity Time Dilation. In fact, there are three conflicts and mistakes in the derivation of Velocity Time Dilation. They are discussed as follows:

1. Speed of Light

Einstein's Special Relativity is based on a postulation that the light speed observed by a ground observer is the same as that of an observer at the light source. This conflicts to the principles of Vision of Light and Photon Inertia Transformation that the light speed changes with observers moving at different speeds and directions with respect to the light source (Fig. 23) [34]. More specifically, it against the Equation of Light Speed that the speed of light is a vector summation of the Absolute Light Speed (3 x 10^8 m/s) and the Inertia Light Speed (the moving speed of light source away from the observer) [35].

Fig. 30 shows a typical example of Einstein's Special Relativity, in which a light clock emits photons to a mirror on the roof of a train while it is moving away from a ground observer.

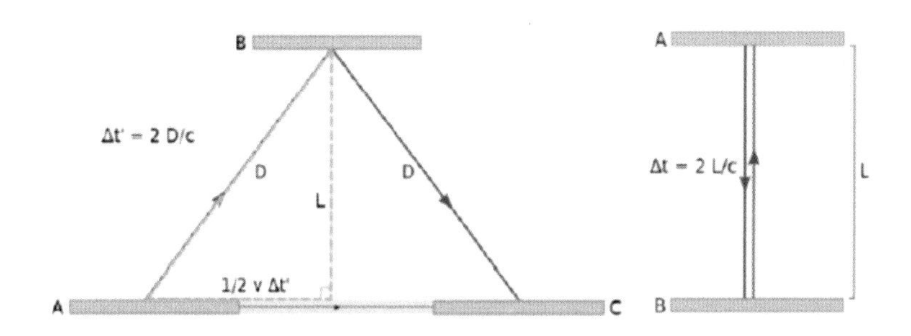

Fig. 30 Moving light clock and Velocity Time Dilation Theory.

Because, Einstein's Special Relativity assumes that light speed is always constant no matter the light sources and observers.

C = C'

$$(\Delta t' C/2)^2 = (\Delta t\, C/2)^2 + (V\, \Delta t'/2)^2$$

Therefore,

$$\Delta t' = (1 - V^2/C^2)^{-1/2}\, \Delta t$$

Because

$$V < C$$

Therefore

$$\Delta t < \Delta t'$$

Where Δt is the light traveling time measured at the light source and $\Delta t'$ is the light traveling time measured on ground.

As a result, the light traveling time measured at the light source Δt is smaller than the light traveling time measured on ground $\Delta t'$. This phenomenon is called Velocity Time Dilation.

However, according to Vision of Light and Photon Inertia Transformation, light speed is not always constant. In Fig. 30, D is the Vision of Light observed on the ground (reference point) and L is the Vision of Light observed at the light source. Also for the same event, time duration is the same, no matter the observation. Therefore,

$$C' = 2D/\Delta t' \qquad \& \qquad C = 2L/\Delta t$$

$$D^2 = L^2 + (V\Delta t'/2)^2$$

$$\Delta t' = \Delta t$$

$$(C'\Delta t/2)^2 = (C\Delta t/2)^2 + (V\Delta t/2)^2$$

And

$$C' = (C^2 + V^2)^{1/2}$$

The light speed observed on the ground is $C' = (C^2 + V^2)^{1/2}$ which is larger than that observed at light source $C = 3x10^8$ m/s.

As a result, oppose to Einstein's Special Relativity and Velocity Time Dilation, light speed is not constant and time doesn't change with velocity neither.

2. Direction of Light

What if the light clock is placed in a tilted angle or horizontal direction instead of a perpendicular direction, with respect to the train moving direction (Fig. 30), do we still have the same Velocity Time Dilation? The answer is no.

Because only for vertical triangle, $\Delta t' = (1 - V^2/C^2)^{-1/2} \Delta t$ works, otherwise this formula is not applicable if the light clock sits at a tilted angle or horizontal position with respect to the moving direction.

3. Twin Paradox

Motion is relative. Whatever the motion twin brothers experienced in their own time system, either in a spaceship or on earth, they are identical except in opposite directions. Slower time and younger age can be claimed by both brothers, which conflicts with the common principles of logical thinking. Twin Paradox [45] proves that Velocity Time Dilation is a false theory and can't never exist.

Furthermore, a similar conflicts are also shown in the derivation of Lorentz transformation, where $\Delta t' = (1 - V^2/C^2)^{-1/2} \Delta t$ is obtained by the postulations that the speed of light C is a constant and the reference system is moving only in the horizontal direction (X direction).

9.13. Einstein's General Relativity

Einstein's General Relativity [39] [46] claimed that acceleration is the principle factor in the universe. Because time can be influenced by acceleration and acceleration can be changed by gravitational force, therefore, clocks that is far from massive bodies, or at higher gravitational potential, run more quickly and clocks close to massive bodies or at lower gravitational potential run more slowly. This phenomenon is known as "Gravitational Time Dilation" [47].

Time is measured by the cycles of a fundamental process in the universe. Since Wu's Pairs are the building blocks of all matter, its circulation cycle is the natural clock of all processes and events. According to Wu's Yangton and Yington Theory, a large gravitational field will cause the increase of the period and decrease the frequency of Yangton and Yington circulation, and thus slow down the cycles of the clock (time). This agrees with Einstein's General Relativity and Gravitational Time Dilation.

However, because acceleration is dependent on the total force, and gravitational force is only one of the Four Basic Forces, Einstein's general relativity is true only when acceleration is solely caused by the gravitational force. In addition, Einstein missed totally the influence caused by the aging of the universe, all because that he has absolutely no idea about Wu's Pairs and Yangton and Yington Theory in his time around 1910s.

Chapter Ten

Spacetime

What Is Spacetime?

How Does Spacetime Change with Gravity?

10.1. Definition of Spacetime

In the universe, the length and time of an object or event can be measured and represented by a four dimensional Spacetime System [x, y, z, t](l_s, t_s) at a reference point. In which [x, y, z] are the position coordinates representing the amounts of unit length (l_s) on three perpendicular axes measured at the reference point (Cartesian coordinate system) and [t] is the time coordinate representing the amount of unit time (t_s) measured at the reference point. For the same object and event, their lengths and durations do not change with the Spacetime or any other measurement method. However, subject to different Spacetime systems, different amounts of lengths and durations can be measured by different unit lengths and unit times. In "Normal Spacetime", Normal Unit Length and Normal Unit Time are independent to each other, such as "meter" and "second" in the MKS system.

10.2. Wu's Spacetime

Wu's Spacetime [x, y, z, t](l_{yy}, t_{yy}) [3] is a special four dimensional system that is defined by the Wu's Unit Length l_{yy} (the diameter of Wu's Pairs) and Wu's Unit Time t_{yy} (the period of Wu's Pairs) at the reference point. Both Wu's Unit Length and Wu's Unit Time are dependent on the gravitational field and aging of the universe at the reference point. Also, they are correlated to each other by Wu's Spacetime Theory ($t_{yy} = \gamma l_{yy}^{3/2}$) [41].

10.3. Einstein's Spacetime

Einstein's Spacetime is relative and inextricably interwoven into what has become known as the space-time continuum. Unlike the Normal Spacetime and Wu's Spacetime, Einstein's Spacetime is not a reference system. Instead, it is a solution of Einstein's Field Equations with a four dimensional space-time continuum derived from a nonlinear geometry system to a Normal Spacetime System on earth. Reflecting the distribution of matter and energy, the derivative of the curvature of the space-time continuum represents the Amount of Normal Unit Acceleration in a Normal Spacetime System on earth.

10.4. Wu's Spacetime Versus Einstein's Spacetime

In Normal Spacetime, Normal Unit Length (l_s) and Normal Unit Time (t_s) are independent to each other. However, in Wu's Spacetime, both Wu's Unit Length (l_{yy}) and Wu's Unit Time (t_{yy}) are dependent on the gravitational field and aging of the universe at the reference point. They are also correlated to each other by Wu's Spacetime Theory ($t_{yy} = \gamma l_{yy}^{3/2}$).

On the other hand, Einstein's Spacetime is a solution of Einstein's Field Equations with a four dimensional space-time continuum derived from a nonlinear geometry system to a Normal Spacetime System on earth. Reflecting the distribution of matter and energy, the derivative of the curvature of the space-time continuum represents the Amount of Normal Unit Acceleration in a Normal Spacetime System on earth. In contrast, Wu's Spacetime on earth [x, y, z, t](l_{yy0}, t_{yy0}) is a four dimensional system with coordinates based on Wu's Unit Length l_{yy0} and Wu's Unit Time t_{yy0} on earth which are related to each other by $t_{yy0} = \gamma l_{yy0}^{3/2}$.

10.5. Distribution of Wu's Unit Length

According to Yangton and Yington Theory, at any point in space, Wu's Unit Length l_{yy} and Wu's Unit Time t_{yy} ($t_{yy} = l_{yy}^{3/2}$) are dependent on the local gravitational field. Therefore, a two dimensional coordination matrix composed of Wu's Unit Squares (1 Wu's Unit Length x 1 Wu's Unit Length) can be used as a map to reflect the distribution of Wu's Unit Length in an object caused by the gravitational field (Fig. I). In other words, the gravitational field can reflect the distribution of Wu's Unit Length and Wu's Unit Time as well as the shape and motion of the corresponding identical object and event.

In a large gravitational field, as a three dimensional coordination matrix composed of Wu's Unit Cubes (1 Wu's Unit Length x 1 Wu's Unit Length x 1 Wu's Unit Length), the structure of an object is expanded and matter is depleted from the matrix. As a result, a Black Hole made of a hollow structure with a singularity composed of a high density core in the center can be predicted.

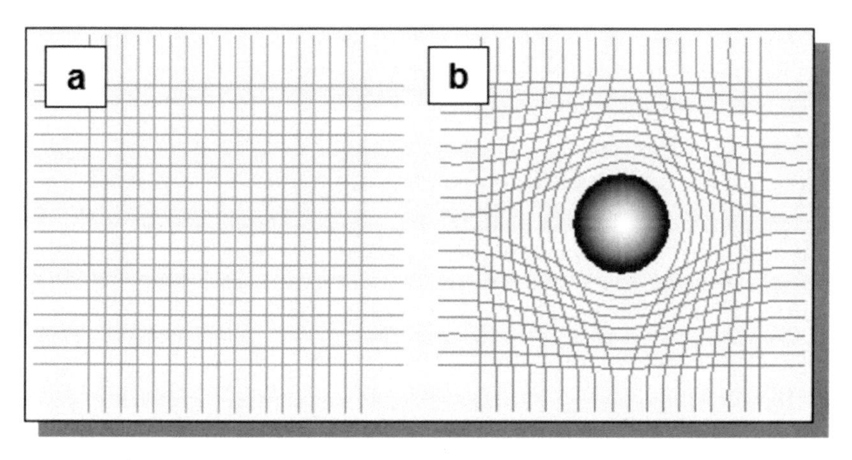

Fig. I (a) A coordination matrix in a homogeneous gravitational field (b) The same coordination matrix in an inhomogeneous field with a big massive core in the center.

10.6. Wu's Spacetime Theory

The period (t_{yy}) and the size (l_{yy}) of the circulation orbit of Wu's Pairs (Fig. 21) are correlated to each other as follows:

Because

$V^2r = K$

$T = 2\pi r/V$

$T^2 = 4\pi^2 r^2/V^2 = 4\pi^2 r^3/V^2 r = 4\pi^2 r^3/K$

$T = 2\pi K^{-1/2} r^{3/2} = \pi (2K)^{-1/2} d^{3/2}$

Given

$$\gamma = \pi (2K)^{-1/2}$$

Therefore,

$$t_{yy} = \gamma l_{yy}^{3/2}$$

Where K is Wu constant, t_{yy} is the circulation period (T) of Wu's Pairs called "Wu's Unit Time", l_{yy} is the size of the circulation orbit (2r = d) of Wu's Pairs called "Wu's Unit Length", and γ is a constant called Wu's Spacetime constant. This is named "Wu's Spacetime Theory" [41].

10.7. Velocity and Spacetime

Because of "Wu's Spacetime Theory",

$t_{yy} = \gamma l_{yy}^{3/2}$

Therefore,

$$l_{yy}/t_{yy} = \gamma^{-1} l_{yy}^{-1/2}$$

For a moving object,

$V = v (l_s/t_s)$

$l_s = m l_{yy}$

$t_s = n t_{yy}$

$V = v (m/n)(l_{yy}/t_{yy})$

Therefore,

$$V = v\ m\ n^{-1}\ \gamma^{-1}\ l_{yy}^{-1/2}$$

Where V is the velocity, "v" is the Amount of Normal Unit Velocity, γ is the Wu's Spacetime constant, m is the constant of Normal Unit Length, n is the constant of Normal Unit Time and l_{yy} is Wu's Unit Length.

For a corresponding identical motion, the Amount of Normal Unit Velocity "v" is a constant, therefore the velocity V is proportional to $l_{yy}^{-1/2}$.

Since the emission of photon from a substance is a process of corresponding identical event, the Amount of Absolute Light Speed 3×10^8 maintains a constant, therefore the Absolute Light Speed C is proportional to $l_{yy}^{-1/2}$ [41].

$$C = c\ m\ n^{-1}\ \gamma^{-1}\ l_{yy}^{-1/2}$$

$$c = 3 \times 10^8$$

$$C \infty\ l_{yy}^{-1/2}$$

As a result, for photon emission at high gravitational field or in ancient universe, because the size of Wu's Pair l_{yy} is bigger and the Amount of Absolute Light Speed "c" is a constant (3×10^8), therefore the Absolute Light Speed C is slower.

However, for photon emission from the same light source, no matter of the frequency, the Absolute Light Speed "C" always maintains constant 3×10^8 cm/s observed at the light source (where cm/s is a Normal Unit Velocity on the light source).

10.8. Deflection of Light

The first observation of light deflection was performed by Arthur Eddington and his collaborators during the total solar eclipse of May 29, 1919 [78] when the stars near the Sun (at that time in the constellation Taurus) could be observed. Starlight that passes close to the sun before reaching us gets deflected (Fig. J).

This starlight will thus reach us from a slightly different direction than when the sun is in some different region of the sky. Accordingly, the star's position in the night sky is shifted slightly.

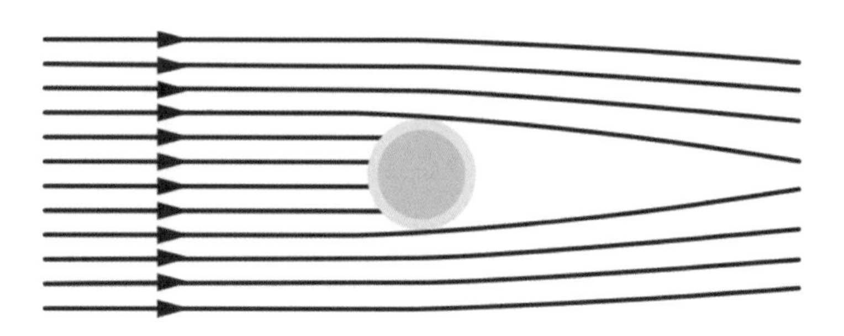

Fig. J Light deflection in the gravitational field of the sun.

In the early 20th century, Einstein successfully explained this phenomenon by his general relativity theory. He claimed that because space-time is highly curved around heavy mass, light rays can thus be deflected when passing by.

One important application of the light deflection effect is "Gravitational Lensing", in which two or more images of one far-away object can be observed (Fig. K).

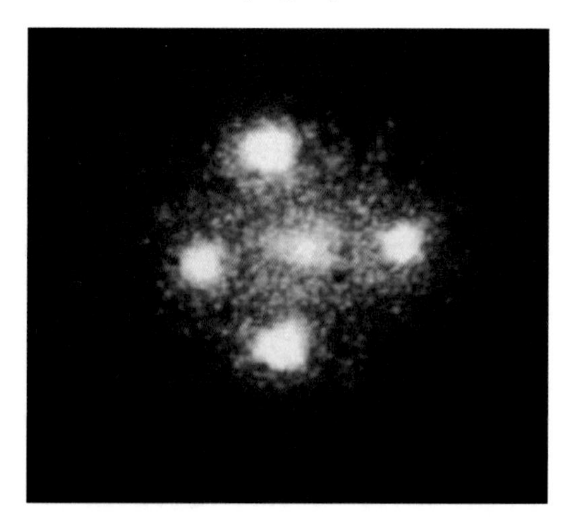

Fig. K Gravitational lenses generate an Einstein cross, image of the Hubble Space Telescope © NASA/ESA/STScI.

Masses acting as gravitational lenses have now become a standard tool of astronomy. They allow astronomers to infer the masses of cosmic objects, and the structure and size scale of the universe (with some caveats). Through their magnifying effect, gravitational lenses have also been used to observe the properties of very distant galaxies and quasars, as well as to search for planets around distant stars.

Light deflection can also be explained by Yangton and Yington Theory. Since the motion of photon is considered as a process of corresponding identical event, the Amount of Absolute Light Speed 3×10^8 maintains a constant, therefore the Absolute Light Speed C is proportional to $l_{yy}^{-1/2}$ [41].

$$C = c\, m\, n^{-1}\, \gamma^{-1}\, l_{yy}^{-1/2}$$

$$c = 3 \times 10^8$$

$$C \propto l_{yy}^{-1/2}$$

When a light beam travels close to a massive star, gravitational field becomes extremely large which makes Wu's Unit Length l_{yy} bigger and Absolute Light Speed C smaller ($C \propto l_{yy}^{-1/2}$). As a result, the path of the light beam bends toward the star in order to maintain the same frequency and the coherency while decreasing the light speed [80] [Annex 31]. This phenomenon is called "Deflection of Light".

10.9. Perihelion Precession of Mercury

A long-standing problem in the study of the Solar System was that the orbit of Mercury did not behave as required by Newton's equations. In fact, it is found that the point of closest approach (Perihelion) of Mercury to the sun does not always occur at the same place but that it slowly moves around the sun (Fig. L). This rotation of the orbit is called a "Perihelion Precession" [79].

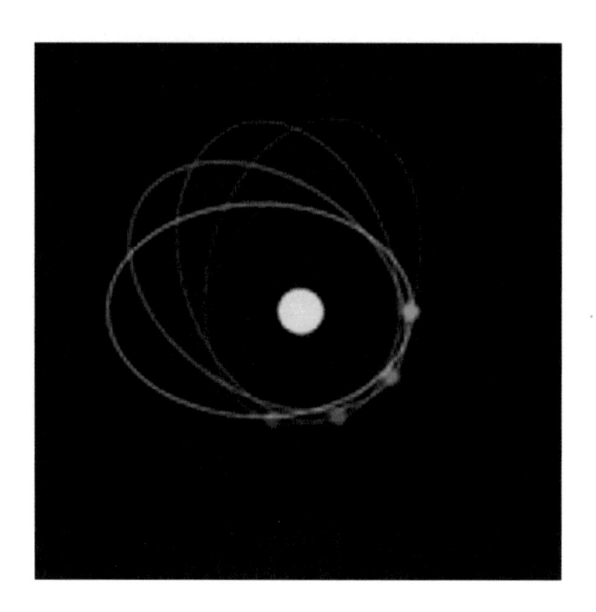

Fig. L Artist's version of the precession of Mercury's orbit.

As seen from Earth the precession of Mercury's orbit is measured to be 5600 seconds of arc per century (one second of arc=1/3600 degrees). Newton's equations, taking into account all the effects from the other planets (as well as a very slight deformation of the sun due to its rotation) and the fact that the Earth is not an inertial frame of reference, predicts a precession of 5557 seconds of arc per century. There is a discrepancy of 43 seconds of arc per century.

This discrepancy cannot be accounted for using Newton's formalism. Many ad-hoc fixes were devised (such as assuming there was a certain amount of dust between the Sun and Mercury) but none were consistent with other observations. In contrast, Einstein was able to predict, without any adjustments whatsoever, that the orbit of Mercury should precess by an extra 43 seconds of arc per century.

In a curved spacetime a planet does not orbit the Sun in a static elliptical orbit, as in Newton's theory. Rather, the orbit is obliged to precess because of the curvature of spacetime. When Einstein calculated the magnitude of this effect for Mercury he got precisely the previously unexplained 43''. He correctly took the view that this was an important confirmation of his general relativity theory.

Perihelion Precession of Mercury can also be interpreted by Yangton and Yington Theory as follows:

Because

$$V = v\, m\, n^{-1}\, \gamma^{-1}\, l_{yy}^{-1/2}$$

Where V is the velocity, “v” is the Amount of Normal Unit Velocity, γ is the Wu’s Spacetime constant, m is the constant of Normal Unit Length, n is the constant of Normal Unit Time and l_{yy} is Wu’s Unit Length.

According to Wu’s Spacetime Theory, for a corresponding identical motion such as that Mercury circulating around the sun, the Amount of Normal Unit Velocity “v” is a constant, and the velocity “V” is proportional to $l_{yy}^{-1/2}$. When Mercury moves close to the sun, gravitational field becomes extremely large which makes l_{yy} bigger and V smaller. As a result, in order to maintain the structure coherency, Perihelion of Mercury precesses to compensate the changes of the circulation speed in each orbiting cycle caused by the extremely high gravitational field [80][Annex 32]. This phenomenon is called Perihelion Precession of Mercury.

10.10. Photon and Spacetime

For a photon moving in space,

$\nu = 1/t_{yy}$

Because of “Wu’s Spacetime Theory”,

$t_{yy} = \gamma l_{yy}^{3/2}$

Therefore,

$$\nu \propto l_{yy}^{-3/2}$$

Because,

$$C \propto l_{yy}^{-1/2}$$

$\lambda = C/\nu$

Therefore,

$$\lambda \propto l_{yy}$$

When the universe grows older, the circulation orbit (2r) of the Wu's Pair becomes smaller. Since V^2r is always a constant ($V^2r = k$) for an inter-attractive circulating pair such as a Wu's Pair (Fig. 21), the circulation speed (V) of a Wu's Pair becomes faster. Also, the circulation period ($T = 2\pi r/V$) of the Wu's Pair gets shorter. In other words, Wu's Unit Time ($t_{yy} = T$) and Wu's Unit Length ($l_{yy} = 2r$) both become smaller. As a result, when the universe grows older, the frequency ($\nu = 1/T$) of a photon becomes bigger, the light speed ($C \propto l_{yy}^{-1/2}$) becomes faster, and the wavelength ($\lambda \propto l_{yy}$) becomes smaller.

For a high gravitational field, the circulation speed (V) of a Wu's Pair becomes slower. Since V^2r is always a constant ($V^2r = k$) for an inter-attractive circulating pair such as a Wu's Pair, the size of circulation orbit (2r) of Wu's Pair becomes larger. Also, the circulation period ($T = 2\pi r/V$) of Wu's Pair gets longer. In other words, Wu's Unit Time ($t_{yy} = T$) and Wu's Unit Length ($l_{yy} = 2r$) both become bigger. As a result, for a low gravitational field, the frequency ($\nu = 1/T$) of a photon becomes smaller, the light speed ($C \propto l_{yy}^{-1/2}$) becomes slower, and the wavelength ($\lambda \propto l_{yy}$) becomes larger.

Furthermore, both Wu's Unit Length l_{yy} and Wu's Unit Time t_{yy} can be represented by Absolute Light Speed C as follows:

Because

$$C \propto l_{yy}^{-1/2}$$

$$t_{yy} \propto l_{yy}^{3/2}$$

Therefore,

$$l_{yy} \propto 1/C^2$$

$$t_{yy} \propto 1/C^3$$

As a result, photon can be considered as a marker of the Spacetime at the original light source. The photon's frequency ($\nu = 1/t_{yy}$), Absolute Light Speed ($C \propto l_{yy}^{-1/2}$) and wavelength ($\lambda \propto l_{yy}$) can carry the information of l_{yy} and t_{yy} of the Spacetime at the original light source deep into the universe. In other words, photon bears the DNA of the original light source.

10.11. Acceleration and Spacetime

Because of "Wu's Spacetime Theory",

$t_{yy} = \gamma l_{yy}^{3/2}$

Therefore,

$l_{yy}/t_{yy}^2 = \gamma^{-2} l_{yy}^{-2}$

For an accelerating object,

$A = a\ (l_s/t_s^2)$

$l_s = m l_{yy}$

$t_s = n t_{yy}$

$A = a\ (m/n^2)(l_{yy}/t_{yy}^2)$

Therefore,

$$A = a\ m\ n^{-2}\ \gamma^{-2}\ l_{yy}^{-2}$$

Where A is the acceleration, a is the Amount of Normal Unit Acceleration, γ is the Wu's Spacetime constant, m is the constant of Normal Unit Length, n is the constant of Normal Unit Time and l_{yy} is Wu's Unit Length.

For a corresponding identical acceleration, the Amount of Normal Unit Acceleration "a" is a constant, therefore,

$$A \infty l_{yy}^{-2}$$

As a result, for a corresponding identical acceleration at high gravitational field or in ancient universe, because the size (l_{yy}) of Wu's Pair is bigger, therefore the acceleration is slower.

10.12. Same Object and Event Observed at Different Reference Points

For the same object and event observed at different reference points, the Amount of Unit Length, Amount of Unit Time, Amount of Unit Velocity and Amount of Unit Acceleration change with the Unit Length, Unit Time, Unit Velocity and Unit Acceleration as follows:

A. Length

The length L of an object or event can be measured by the Normal Unit Length (such as meter).

$L = l\, l_s$

Where l is the Amount of Normal Unit Length and l_s is the Normal Unit Length.

And

$l_s = m\, l_{yy}$

Where m is a constant, l_s is Normal Unit Length and l_{yy} is Wu's Unit Length.

Therefore,

$$L = l\, m\, l_{yy}$$

For the same object and event, L is a constant. Therefore,

$$l \propto l_{yy}^{-1}$$

For an object or event on a massive star, because of the smaller l_{yy0} on earth, bigger Amount of Normal Unit Length l_0 can be observed on earth [72].

B. Time

The time T of an object or event can be measured by the Normal Unit Time (such as second).

$T = t\, t_s$

Where t is the Amount of Normal Unit Time and t_s is the Normal Unit Time.

And

$t_s = n\, t_{yy}$

Where n is a constant, t_s is Normal Unit Time and t_{yy} is Wu's Unit Time.

Therefore,

$T = t\, n\, t_{yy}$

Also because of Wu's Spacetime Theory [41],

$t_{yy} = \gamma\, l_{yy}^{3/2}$

Where γ is Wu's Spacetime Constant.

Therefore,

$$T = t\, n\, \gamma\, l_{yy}^{3/2}$$

For the same object and event, T is a constant. Therefore,

$$t \propto l_{yy}^{-3/2}$$

For an object or event on a massive star, because of the smaller l_{yy0} on earth, bigger Amount of Normal Unit Time t_0 can be observed on earth [72].

C. Velocity

The velocity V of an object or event can be measured by the Normal Unit Velocity (such as m/s).

$V = v\, (l_s/t_s)$

Where v is the Amount of Normal Unit Velocity and l_s/t_s is the Normal Unit Velocity.

Because

$$V = v\, m\, n^{-1}\, \gamma^{-1}\, l_{yy}^{-1/2}$$

Where γ is Wu's Spacetime constant, m is the constant of Normal Unit Length and n is the constant of Normal Unit Time.

For the same object and event, V is a constant. Therefore,

$$v \propto l_{yy}^{1/2}$$

For an object or event on a massive star, because of the smaller l_{yy0} on earth, smaller Amount of Normal Unit Velocity v_0 can be observed on earth [72].

D. Acceleration

The acceleration A of an object or event can be measured by the Normal Unit Acceleration (such as m/s^2).

$A = a\ (l_s/t_s^2)$

Where a is the Amount of Normal Unit Acceleration and l_s/t_s^2 is the Normal Unit Acceleration.

Because

$$A = a\ m\ n^{-2}\ \gamma^{-2}\ l_{yy}^{-2}$$

Where γ is Wu's Spacetime constant, m is the constant of Normal Unit Length and n is the constant of Normal Unit Time.

For the same object and event, A is a constant. Therefore,

$$a \infty l_{yy}^{2}$$

For an object or event on a massive star, because of the smaller l_{yy0} on earth, smaller Amount of Normal Unit Acceleration a_0 can be observed on earth [72].

10.13. Corresponding Identical Object and Event Observed on Earth

When a corresponding identical object or event on a massive star is observed on earth, the total length, time, velocity and acceleration change with the Corresponding Identical Unit Quantities at the star:

A. Length

Because

$$L = l\ m\ l_{yy}$$

For a corresponding identical object or event, l is a constant. Therefore,

$$L \infty\ l_{yy}$$

For a corresponding identical object or event on a massive star, because of the large gravitational force, both l_{yy} and L are bigger than that on earth. Furthermore, to observe the object and event from earth, because of the small Normal Unit Length on earth, big Amount of Normal Unit Length can be observed on earth. As a result, for observation on earth, the Amount of Normal Unit Length of a corresponding identical object or event on a massive star is bigger compared to that of the corresponding identical object or event on earth [72].

B. Time

Because

$$T = t\ n\ \gamma\ l_{yy}^{3/2}$$

For a corresponding identical object or event, t is a constant. Therefore,

$$T \infty\ l_{yy}^{3/2}$$

For a corresponding identical object or event on a massive star, because of the large gravitational force, both l_{yy} and T are bigger than that on earth. Furthermore, to observe the object and event from earth, because of the small Normal Unit Time on earth, big Amount of Normal Unit Time can be observed on earth. As a result, for observation on earth, the Amount of Normal Unit Time of a corresponding identical object or event on a massive star is bigger compared to that of the corresponding identical object or event on earth [72].

C. Velocity

Because

$$V = v\, m\, n^{-1}\, \gamma^{-1}\, l_{yy}^{-1/2}$$

For a corresponding identical object or event, v is a constant. Therefore,

$$V \infty\, l_{yy}^{-1/2}$$

For a corresponding identical object or event on a massive star, because of the large gravitational force, l_{yy} is bigger but V is smaller than that on earth. Furthermore, to observe the object and event from earth, because of the large Normal Unit Velocity ($l_{yy}/t_{yy} \infty\, l_{yy}^{-1/2}$) on earth, small Amount of Normal Unit Velocity can be observed on earth. As a result, for observation on earth, the Amount of Normal Unit Velocity of a corresponding identical object or event on a massive star is smaller compared to that of the corresponding identical object or event on earth [72].

D. Acceleration

Because

$$A = a\, m\, n^{-2}\, \gamma^{-2}\, l_{yy}^{-2}$$

For a corresponding identical object or event, a is a constant. Therefore,

$$A \infty\, l_{yy}^{-2}$$

For a corresponding identical object or event on a massive star, because of the large gravitational force, l_{yy} is bigger but A is smaller than that on earth. Furthermore, to observe the object and event from earth, because of the large Normal Unit

Acceleration ($l_{yy}/t_{yy}^2 \propto l_{yy}^{-2}$) on earth, small Amount of Normal Unit Acceleration can be observed on earth. As a result, for observation on earth, the Amount of Normal Unit Length of a corresponding identical object or event on a massive star is smaller compared to that of the corresponding identical object or event on earth [72].

In conclusion, according to Yangton and Yington Theory, for a corresponding identical object or event on a massive star under a large gravitational field, its length L, time T are bigger, but velocity V and acceleration A become smaller. Furthermore, while observing on earth, because of the relatively smaller Wu's Unit Length and Time, the same corresponding identical object or event on a massive star shows even larger length and time but smaller velocity and acceleration compared to that of the corresponding identical object or event on earth. This result agrees very well with general relativity.

10.14. Einstein's Field Equations

The Einstein field equations (EFE) may be written in the form:

$$R_{\mu\nu} - \frac{1}{2} R \, g_{\mu\nu} + \Lambda \, g_{\mu\nu} = \frac{8\pi G}{c^4} T_{\mu\nu}$$

where $R_{\mu\nu}$ is the Ricci curvature tensor, R is the scalar curvature, $g_{\mu\nu}$ is the metric tensor, Λ is the cosmological constant, G is Newton's gravitational constant, c is the speed of light in vacuum (a constant), and $T_{\mu\nu}$ is the stress–energy tensor.

The Einstein field equations comprise the set of 10 equations in Albert Einstein's general theory of relativity that describe the fundamental interaction of gravitation as a result of spacetime being curved by mass and energy. First published by Einstein [39] in 1915 as a tensor equation, the EFE relate local spacetime curvature (expressed by the Einstein tensor) with the local energy and momentum within that spacetime (expressed by the stress–energy tensor).

To avoid the universe from collapsing, Einstein added the cosmological constant into the formula to balance the attraction force caused by the gravity. However, after Hubble showed us that the universe is expanding, this term was not longer necessary, because the universe is not static. Einstein later felt that the inclusion of this term was the biggest blunder of his career.

Similar to the way that electromagnetic fields are determined using charges and currents via Maxwell's equations, the EFE are used to determine the spacetime geometry resulting from the presence of mass–energy and linear momentum, that is, they determine the metric tensor of spacetime for a given arrangement of stress–energy in the spacetime. The relationship between the metric tensor and the Einstein tensor allows the EFE to be written as a set of nonlinear partial differential equations when used in this way. The solutions of the EFE are the components of the metric tensor. The inertial trajectories of particles and radiation (geodesics) in the resulting geometry are then calculated using the geodesic equation.

As well as obeying local energy–momentum conservation, the EFE reduce to Newton's law of gravitation where the gravitational field is weak and velocities are much less than the speed of light [56].

Exact solutions for the EFE can only be found under simplifying assumptions such as symmetry. Special classes of exact solutions are most often studied as they model many gravitational phenomena, such as rotating black holes and the expanding universe. Further simplification is achieved in approximating the actual spacetime as flat spacetime with a small deviation, leading to the linearized EFE. These equations are used to study phenomena such as gravitational waves.

10.15. Wu's Spacetime Field Equations

Any object "m" at a distance "R" from a massive star "M" can have an acceleration "A" generated from the gravitational force "F" between the object and the star.

Because of Newton's Second Law of Motion and Newton's Law of Universal Gravitation,

$F = m A$

$F = G m M/R^2$

Therefore,

$$A = GM/R^2$$

Where A is the acceleration, m is the mass of the object, M is the mass of the star and R is the distance between the object and the star. Because GM/R^2 is the gravitational field surrounding the massive star, this equation is called "Field Equation".

According to Yangton and Yington Theory, the acceleration of an object can be measured at a reference point:

$A = a\, m\, n^{-2} \gamma^{-2} l_{yy}^{-2}$

Where A is acceleration, a is the Amount of Normal Unit Acceleration, γ is the Wu's Spacetime constant, m is the constant of Normal Unit Length, n is the constant of Normal Unit Time and l_{yy} is Wu's Unit Length at the reference point.

Because,

$A = GM/R^2$

$A = a\, m\, n^{-2} \gamma^{-2} l_{yy}^{-2}$

Also,

$C \propto l_{yy}^{-1/2}$

$C^{-4} \propto l_{yy}^{2}$

Therefore,

$$a = \sigma\, \gamma^2\, l_{yy}^{2}\, G\, M/R^2$$

$$a = \delta\, \gamma^2\, C^{-4}\, G\, M/R^2$$

These are named "Wu's Spacetime Field Equations" [57]. Where a is the Amount of Normal Unit Acceleration, σ and δ are constants, γ is Wu's Spacetime constant, G is the gravitational constant [Annex 30], l_{yy} is Wu's Unit Length and C is the Absolute Light Speed ($C \propto l_{yy}^{-1/2}$) at the reference point.

Wu's Spacetime Field Equation represents the Amount of Normal Unit Acceleration "a" measured based on Wu's Unit Length l_{yy} at the reference point, which reflects the distribution of energy and momentum of matter. Since Wu's Unit Length l_{yy} is an unknown quantity, Absolute Light Speed ($C \propto l_{yy}^{-1/2}$) which can be measured by redshift is used in Wu's Spacetime Field Equation.

At a position close to a massive spherical mass (star), because distance R is smaller, therefore the Amount of Normal Unit Acceleration "a" (also the curvature) is bigger. As a result, a deeper Space-time continuum can be observed on earth (Fig. D) which predicts the existence of a Black Hole.

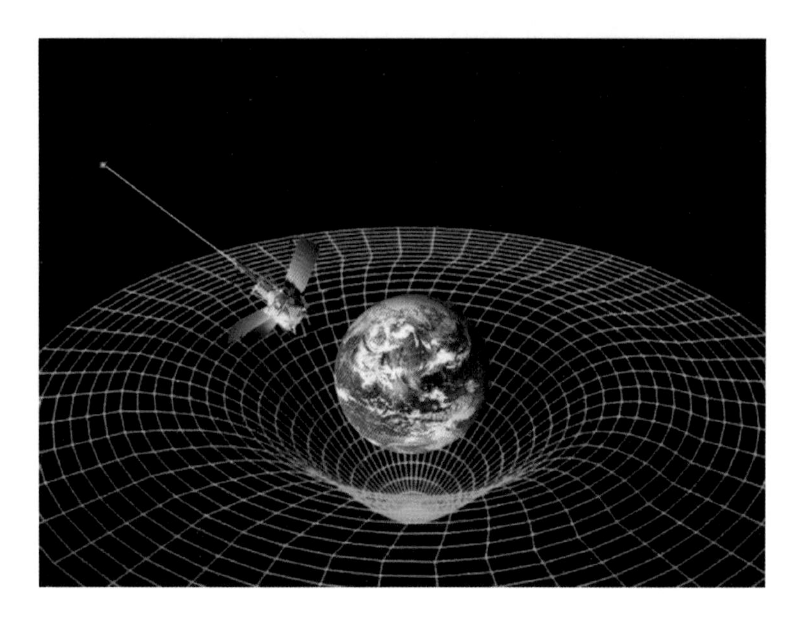

Fig. D. Earth and its surrounding spacetime continuum.

10.16. Wu's Spacetime Field Equations versus Einstein's Field Equations

Acceleration and Wu's Spacetime Field Equation can be represented by Wu's Unit Length l_{yy} and the Absolute Light Speed (C) at the reference point.

$$A = a\ m\ n^{-2}\ \gamma^{-2}\ l_{yy}^{-2}$$

$$a = \delta\ \gamma^2\ C^{-4}\ G\ M/R^2$$

Where "a" is the Amount of Normal Unit Acceleration (the curvature of Wu's Spacetime), δ is a constant, γ is Wu's Spacetime constant and C is the Absolute Light Speed ($C \infty l_{yy}^{-1/2}$) at the reference point.

For the same object and event, acceleration and Wu's Spacetime Field Equation can also be represented by Wu's Unit Length l_{yy0} and the Absolute Light Speed C_0 observed on earth.

$$A = a_0\, m\, n^{-2}\, \gamma^{-2}\, l_{yy0}^{-2}$$

$$a_0 = \delta\, \gamma^2\, C_0^{-4}\, G\, M/R^2$$

Where "a_0" is the Amount of Normal Unit Acceleration measured on earth, δ is a constant, γ is Wu's Spacetime constant, C_0 is the Absolute Light Speed on earth ($3\text{x}10^8$ m/s) and l_{yy0} is Wu's Unit Length on earth.

According to Yangton and Yington Theory, Wu's Unit Length l_{yy} on a massive star is much bigger than l_{yy0} on earth. Because $a \propto C^{-4} \propto l_{yy}^2$, therefore for the same object and event, the Amount of Normal Unit Acceleration "a" measured on the star is much bigger than "a_0"measured on earth. In other words, for a massive star, Wu's Spacetime Field Equation observed on the star has deeper slope (bigger curvature) than Wu's Spacetime Field Equation observed by Wu's Spacetime System on earth (Fig. H) which is equivalent to Einstein's Field Equation observed by Normal Spacetime System on earth.

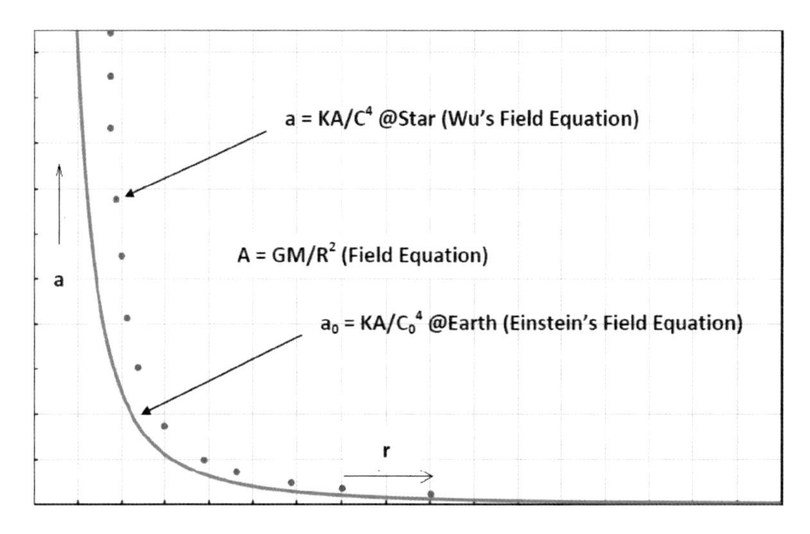

Fig. H Comparison between Einstein's Field Equation (blue solid line) and Wu's Field Equation (red dotted line).

Einstein's Field Equations has a solution with a four dimensional space-time continuum derived from a nonlinear geometry system to a Normal Spacetime System on earth. Reflecting the distribution of matter and energy, the derivative of the curvature of the space-time continuum represents the Amount of Normal Unit Acceleration in a Normal Spacetime System on earth.

In contrast, Wu's Field Equation represents the Amount of Normal Unit Acceleration (curvature) in Wu's Spacetime reflecting the gravitational force and the distribution of matter on earth where Wu's Unit Length l_{yy0} and Wu's Unit Time t_{yy0} are correlated to each other by $t_{yy0} = \gamma l_{yy0}^{3/2}$.

Because the same term GC_0^{-4} appears in both equations, Einstein's Field Equation and Wu's Spacetime Field Equation observed on earth look like equivalent. However, there is no gravitational force in Einstein's Spacetime Field Equation. Acceleration is derived from the curvature of space-time continuum, which reflects the virtual distribution of matter and energy in the universe. On the other hand, in Wu's Spacetime Field Equation, matter does exist, as is the gravitational force. And the acceleration is indeed caused by the gravitational force.

10.17. Wu-Einstein Field Equation

A similar equation to Einsten's Field Equation can also be derived from Wu's Spacetime Field Equation to reflect the correlation between Space-Time and Matter-Energy as follows:

Because,

$F = mA$

$F = GmM/R^2$

$A = GM/R^2$

And

$A = a\ mn^{-2}\gamma^{-2}l_{yy}^{-2}$

$l_{yy}^{-2} \infty C^4$

$l_{yy}^{-2} = eC^4$

$A = a\, mn^{-2}\gamma^{-2}eC^4$

Where “e” is a constant named “Constant of Light”.

Because

$a = m^{-1}n^2\gamma^2\, e^{-1}C^{-4}\, (GM/R^2)$

Given

$k_1 = m^{-1}n^2\gamma^2\, e^{-1}$

Therefore,

$$a = k_1\, C^{-4}\, G\, (M/R^2)$$

Where k_1 is a constant and this equation is named “Wu’s Spacetime Field Equation”.

$dE = F\, dR$

$dE/dR = F$

$F = m_0A$

$dE/dR = m_0A$

$(dE/m_0)/dR = A$

Given

$E_0 = E/m_0$

Therefore,

$dE_0/dR = A$

Where m_0 is the mass of an object and E_0 is the potential energy of a unit mass (1Kg).

Because

$A = a_0\, mn^{-2}\gamma^{-2}eC_0^{4}$

Therefore,

$a_0 = m^{-1}n^{2}\gamma^{2}\, e^{-1}C_0^{-4}\,(dE_0/dR)$

Given

$k_2 = m^{-1}n^{2}\gamma^{2}\, e^{-1}C_0^{-4}$

Therefore,

$a_0 = k_2\,(dE_0/dR)$

Also,

$a_0 = k_1\, C_0^{-4}\, G\,(M/R^2)$

Therefore,

$k_2\, R^2\,(dE_0/dR) = k_1\, C_0^{-4}\, G\, M$

And

$dR^{-1} = -1R^{-2}dR$

$\kappa = R^{-1}$

Therefore,

$$(dE_0/d\kappa) = -(k_1/k_2)\, C_0^{-4}\, G\, M$$

Where a_0 is Amount of Unit Acceleration and C_0 is Absolute Light Speed on earth. K_1 and K_2 are constants and κ is the curvature of a space-time continuum E_0 (Energy Field). This equation is named “Wu-Einstein Field Equation” [81].

Compare Wu-Einstein Spacetime Field Equation to Einstein’s Field Equation,

$$R_{\mu\nu} - \frac{1}{2} R\, g_{\mu\nu} + \Lambda\, g_{\mu\nu} = \frac{8\pi G}{c^4} T_{\mu\nu}$$

The left hand side of Einstein Field Equation is equivalent to (dE/dκ) which is a geometric term related to the curvature κ of a space-time continuum E [81]. The right hand side of the equation is equivalent to $-(k_1/k_2)C^{-4}$ GM which is a term related to mass, energy and momentum of the object.

10.18. Wu's Spacetime Field Equation and Concentration of Gravitons

According to Wu's Yangton and Yington Theory, both Wu's Unit Time (t_{yy}) and Wu's Unit Length (l_{yy}) are functions of the gravitational field (F_g). Because gravitational field is generated by Graviton Radiation and Contact Interaction [17], it is a function of the concentration of Gravitons ($C_{Graviton}$), therefore, Spacetime [x, y, z, t] (t_{yy}, l_{yy}) can also be represented by the gravitational field [x, y, z, t] (F_g) and concentration of Gravitons field [x, y, z, t] ($C_{Graviton}$) at the reference point.

Therefore, similar to Wu's Spacetime Field Equation, A Spacetime Graviton Concentration Field Equation can be derived as follows:

Because

$\Sigma(M/R^2) \propto (C_{Graviton})$

Therefore,

$$a = \eta\, \gamma^2\, C^{-4}\, G\, (C_{Graviton})$$

This is named "Spacetime Graviton Concentration Field Equation". Where a is the Amount of Normal Unit Acceleration, η is a constant, γ is Wu's Spacetime constant, G is the gravitational constant, C is the speed of light which is a function of Wu's Unit Length l_{yy} that is dependent on the gravitational field and aging of the universe at the reference point, and $C_{Graviton}$ is the concentration of Gravitons.

Spacetime Graviton Concentration Field Equation shows the relationship between the Amount of Normal Unit Acceleration (curvature) of Wu's Spacetime and the distribution of the concentration of Graviton. Therefore, it can be considered as the backbone of Quantum Field Theory.

10.19. Spacetime and Aging of the Universe – Cosmological Redshift

When the universe was young, the circulation orbit (2r) of Wu's Pair was bigger. Since V^2r is always a constant ($V^2r = k$) for an inter-attractive circulating pair such as Wu's Pair, the circulation speed (V) of Wu's Pairs was slower. The circulation period ($T = 2\pi r/V$) of Wu's Pairs was also bigger. In other words, when the universe was young, both Wu's Unit Length ($l_{yy} = 2r$) and Wu's Unit Time ($t_{yy} = T$) were bigger, which means the length was longer, time ran slower, and velocity was slower compared to that on earth today. As a result, light coming from a star greater than 5 billion years ago (5 billion light years away), travels at a slower speed ($C \propto l_{yy}^{-1/2}$) with lower frequency ($\nu = 1/T$) and a larger wavelength ($\lambda \propto l_{yy}$). This phenomenon is known as "Cosmological Redshift" [41] [42].

Because of the shrinkage of Wu's Spacetime with the aging of the universe, Wu's Spacetime Reverse Expansion Theory [52] can be derived to explain Hubble's Law and the expansion of the universe without the Dark Energy for acceleration neither the Cosmological Constant for Field Equations.

10.20. Spacetime and Gravitational Field – Gravitational Redshift

When a gravitational field increases, the attractive force between gravitons also increases. Thus the circulation speed (V) of a Wu's Pair becomes slower. Since V^2r is always a constant ($V^2r = k$) for an inter-attractive circulating pair such as a Wu's Pair, the size of the circulation orbit (2r) of Wu's Pair gets bigger. And the circulation period ($T = 2\pi r/V$) of Wu's Pair also gets bigger. In other words, when the gravitational field increases, both Wu's Unit Length ($l_{yy} = 2r$) and Wu's Unit Time ($t_{yy} = T$) become greater, meaning time runs more slowly, length is longer and velocity is slower compared to that on earth. As a result, light comes from a large gravitational field traveling at a slower speed with a lower frequency and a larger wavelength. This phenomenon is known as "Gravitational Redshift" [40].

Chapter Eleven

Expansion of the Universe

Does the Universe Really Expand And Accelerate?

Does Dark Energy Really Exist?

12.1. Expansion and Acceleration of the Universe

A Redshift [29] [40] occurs whenever a light source moves away from an observer. Another kind of Redshift is the Cosmological Redshift [42], which is due to the expansion of the universe. Sufficiently distant light sources (generally more than a few billion light years away) show the Redshift corresponding to the rate of increase in their distance from earth. It is an intrinsic expansion whereby the scale of space itself changes. Finally, the gravitational Redshift is a relativistic effect observed in electromagnetic radiation moving out of gravitational fields. In the early part of the twentieth century, Slipher [49], Hubble and others made the first measurements of the Redshifts and Blue Shifts of the galaxies beyond the Milky Way. They initially interpreted these Shifts due solely to the Doppler Effect. Later Hubble discovered a rough correlation between the increasing Redshifts and the increasing distance of galaxies. Theorists immediately realized that these observations could be explained by a different mechanism for producing Redshifts. Hubble's Law [50] of the correlation between Redshifts and distances is required by models of cosmology derived from general relativity that have a metric expansion of space. As a result, photons propagating through the expanding space are stretched, creating the cosmological Redshift. According to measurements, the universe's expansion rate was decelerating until about 5 billion years ago due to the gravitational attraction of the matter content of the universe. After that time, the expansion began accelerating. An external energy known as "Dark Energy" is proposed as the reason causing the acceleration of the universe's expansion [49] [51]. However, where Dark Energy originated remains a mystery.

12.2. Dark Energy

By fitting a theoretical model of the composition of the universe to the combined set of cosmological observations, scientists have come up with the composition of about 68% Dark Energy, 27% Dark Matter and 5% normal matter. Dark Matter doesn't emit photons. It is invisible and thus named as Dark Matter. Dark Matter works like glue. Its mass generates sufficient gravity to keep galaxies from drifting apart by spinning. Dark Energy [51] is proposed by scientists for the purpose of explaining the Accelerating Expansion of Universe. Does Dark Energy really exist? Or is it simply imagination? So far, nobody has a clue.

12.3. Hubble's Law

The discovery of the linear relationship between Redshift and distance for stars more than 5 billion years away, coupled with a supposed linear relation between recessional velocity and Redshift yields a straight forward mathematical expression for "Hubble's Law" (Fig. 31) [50] as follows:

$$V = H_0 D$$

Where

- V is the recessional velocity, typically expressed in km/s.
- H_0 is Hubble constant and corresponds to the value of H (often termed the Hubble parameter a value that is time dependent and can be expressed in terms of the scale factor) in the Friedmann equations

- Taken at the time of observation denoted by the subscript "$_0$". This value is the same throughout the universe for a given comoving time.

- D is the proper distance (which can change over time, unlike the comoving distance, which is constant) from the galaxy to the observer, measured in mega parsecs (Mpc) the 3-space defined by given cosmological time. (Recession velocity is just V = dD/dt).

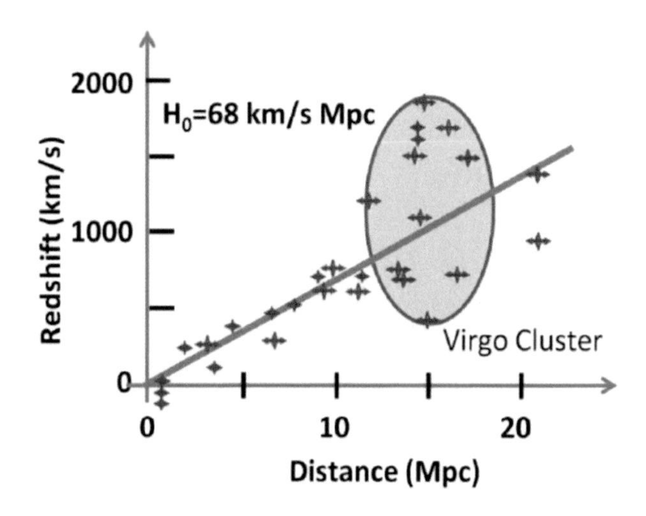

Fig. 31 Hubble's Law – the linear relationship between Redshift and distance.

12.4. Hubble's Law and Acceleration Doppler Effect

Although Hubble's Law is an experimental result, it can be proved by Acceleration Doppler Effect [52]. According to the mathematics in the derivation of Redshift in Acceleration Doppler Effect [52], where a star is moving away from earth at a constant acceleration speed a, D is the distance from the star to earth, P is the distance from the earth to light origin, S is the distance from light source to light origin. Then,

$$D = P - S = Ct + \frac{1}{2} at^2 = (C + \frac{1}{2} at)\, t$$

For stars more than 5 billion years away, the acceleration ½ at becomes much bigger than C (in other words, V is much bigger than C). Therefore,

$D/t = \frac{1}{2} at$

Because

$\lambda_1 = D/vt = (Ct + \frac{1}{2} at^2)/vt = (C + \frac{1}{2} at)/v = \lambda + \frac{1}{2} at/v$

$(\lambda_1 - \lambda)/\lambda = (\frac{1}{2} at)/C$

$(\lambda_1 - \lambda)/\lambda \infty at$

Therefore,

$$D/t \infty (\lambda_1 - \lambda)/\lambda$$

Also,

$V = V_0 + at$

$at \gg V_0$

$V = at$

Therefore,

$$V \infty (\lambda_1 - \lambda)/\lambda$$

Where λ_1 is the wavelength of the photon emitted from the star observed on earth and λ is the wavelength of the photon on earth, $(\lambda_1 - \lambda)/\lambda$ is the redshift, V is the velocity of the star moving away from earth and D/t is the proper distance.

Because both V and D/t are proportional to $(\lambda_1 - \lambda)/\lambda$

Therefore,

$$V = kD/t$$

Given

$$H_0 = k/t$$

Then

$$V = H_0D$$

Where k is a constant and H_0 is Hubble Constant (a time dependent constant).

For those stars they separated from earth at the same time, both t and $H_0 = k/t$ are constants and V-D curve becomes a straight line. Also, when the universe gets older, t is bigger, H_0 is smaller, and V-D curve becomes flat with a smaller slope [58]. Furthermore, for those stars more than 5 billion light years away, 1/t becomes smaller and eventually converges to a constant, so as is H_0. As a result, Redshift is proportional to both D and V, which obeys Hubble's Law (Fig. 31).

According to the Acceleration Doppler Effect and Hubble's Law, the universe is not only expanding but also accelerating. This is named "Universe Expansion and Acceleration Theory".

Chapter Twelve

Shrinkage of Spacetime

Why Spacetime Shrinking?

12.1. Wu's Spacetime Shrinkage Theory

According to the Five Principles of the Universe, through the aging of the universe, Wu's Pair is getting smaller and eventually Yangton will recombine with Yington to destroy each other such that everything will go back to Nothing.

As a consequence, Spacetime [x, y, z, t](l_{yy}, t_{yy}) is shrinking because the diameter of Wu's Pair l_{yy} (Wu's Unit Length) is getting smaller due to the aging of the universe, also the period of the circulation of Wu's Pair t_{yy} (Wu's Unit Time) is shrinking at 3/2 power of l_{yy} according to Wu's Spacetime Theory ($t_{yy} = \gamma l_{yy}^{3/2}$). This is named "Wu's Spacetime Shrinkage Theory".

12.2. Wu's Spacetime Shrinkage Rate

Because of "Wu's Spacetime Theory",

$t_{yy} = \gamma l_{yy}^{3/2}$

Therefore,

$dt_{yy} = \gamma(3/2)l_{yy}^{1/2}dl_{yy}$

$dl_{yy}/dt_{yy} = (\gamma(3/2))^{-1}l_{yy}^{-1/2}$

$dl_{yy}/dt_{yy} = nl_{yy}^{-1/2}$

And

$$V_{yy} = nl_{yy}^{-1/2}$$

Where n is a constant.

In comparison to those in the ancient universe, the size (l_{yy}) of Wu's Pair on the present earth is getting smaller ($dl_{yy} < 0$) and the period (t_{yy}) of the Wu's Pair is also getting shorter($dt_{yy} <0$). The Spacetime Shrinkage Rate V_{yy} of the Wu's Pair is positive inversely proportional to the square root of the diameter of Wu's Pair ($l_{yy}^{-1/2}$).

As a result, when the universe grows older, l_{yy} gets smaller and V_{yy} becomes bigger. In other words, for the same dl_{yy} happened at the later stage, dt_{yy} is smaller and the shrinkage of Wu's Unit Time t_{yy} is slowing down.

12.3. Wu's Spacetime Accelerating Shrinkage Rate

Because of Wu's Spacetime Theory,

$$t_{yy} = \gamma l_{yy}^{3/2}$$

Therefore,

$$dt_{yy} = \gamma(3/2)l_{yy}^{1/2}dl_{yy}$$

$$V_{yy} = dl_{yy}/dt_{yy} = (\gamma(3/2))^{-1}l_{yy}^{-1/2}$$

$$dV_{yy} = (\gamma(3/2))^{-1}(-1/2)l_{yy}^{-3/2}dl_{yy}$$

$$dV_{yy}/dt_{yy} = -ml_{yy}^{-2}$$

Therefore,

$$a_{yy} = -ml_{yy}^{-2}$$

Where m is a constant.

In comparison to those in the ancient universe, the size (l_{yy}) of a Wu's Pair on the present earth is getting smaller ($dl_{yy} < 0$), the period (t_{yy}) of the Wu's Pair is getting shorter ($dt_{yy} <0$). With an increasing V_{yy} ($dV_{yy} > 0$) and a decreasing t_{yy} ($dt_{yy} <0$), the Spacetime

Shrinkage Acceleration Rate a_{yy} of the Wu's Pair is negative and inversely proportional to the square of the diameter of Wu's Pair $(-l_{yy}^{-2})$.

As a result, when the universe grows older, l_{yy} gets smaller and a_{yy} becomes negatively bigger. In other words, for the same dl_{yy} or dV_{yy} happened at the later stage, dt_{yy} becomes much smaller and the shrinkage of Wu's Unit Time is gradually slowing down to zero.

During Spacetime Shrinkage, because of the Principle of Correspondence, although t_{yy} and l_{yy} become smaller, the relative amounts of t_w and l_w, t_s and l_s, also t_n and l_n remain unchanged. The speed, frequency and wavelength of the light generated on earth is always measured the same by the stationary observer on earth. But, the universe becomes bigger as measured by the shrinking l_s on earth. In addition, because of the Spacetime Shrinkage, the light from a star in the ancient universe, a few billion light years away, has a lower velocity (Absolute Light Speed C) and lower frequency but longer wavelength observed on earth. This phenomenon is called Cosmological Redshift.

The theory of Accelerating Expansion of the Universe requires a lot of external energy. This is why scientists have imagined the existence of the mysterious Dark Energy in the universe. Unfortunately, Dark Energy never exist. In fact, because the potential energy can be converted to kinetic energy in the shrinking circulation process of Wu's Pair with no need of external energy, therefore the Spacetime Shrinkage Theory gives a better explanation to the phenomenon so called "The Expansion and Acceleration of the Universe" which actually should be renamed as "The Reverse Expansion of Wu's Spacetime".

12.4. Hubble's Law and Wu's Spacetime Shrinkage Theory

Although Hubble's Law can be used to explain the expansion of the universe that is derived successfully from the Acceleration Doppler Effect, it is hard to believe that a star can move faster than light speed also with an acceleration pumped by a mysterious Dark Energy. To avoid these problems, Wu's Spacetime Reverse Expansion Theory based on Wu's Spacetime Shrinkage Theory is proposed to interpret Hubble's Law.

According to Wu's Spacetime Shrinkage Theory, the shrinkage of the circulation period (t_{yy}) and orbital size (l_{yy}) of Wu's Pairs are caused by the aging of the universe. As a consequence, a photon emitted from a star more than 5 billion years ago has a larger wavelength than that on the present earth, which can cause redshift and to fulfill Hubble's Law.

Fig. C shows a schematic diagram of the visions of star on earth. In the beginning (when photon is emitted from the star), the distance between the star and earth X is the multiplication of the Normal Unit Length L_i and the Amount of Normal Unit Length M_i. At the final stage (when the photon reaches the earth), the distance of the star X becomes the multiplication of the Normal Unit Length L_f and the Amount of Normal Unit Length M_f. The distance of the star X stays the same. But the vision of the star D_E moves from initial distance M_iL_f to the final distance M_fL_f observed on earth. Because M_fL_f is much bigger than M_iL_f, D_E is approximately equal to the distance X between the star and earth (Fig. C).

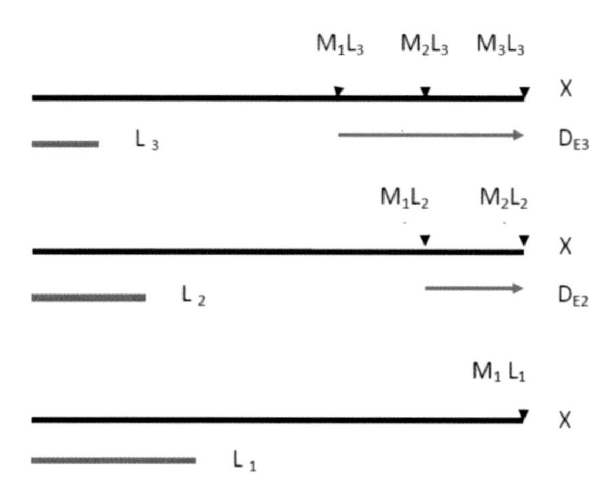

Fig. C The distance of a star measured by a shrinking ruler on earth.

Therefore,

$X = M_f L_f$

$D_E = M_f L_f - M_i L_f = (M_f - M_i) L_f$

$D_E = M_f L_f (M_f - M_i)/M_f$

$$D_E = X (1 - M_i/M_f)$$

Because

$M_i L_i = M_f L_f = X$

$M_i/M_f = L_f/L_i$

Therefore,

$D_E = X (1 - L_f/L_i)$

And

$$D_E = X (L_i - L_f)/L_i$$

Because L is a corresponding identical normal length,

$L \infty l_{yy} \infty \lambda$

$(L_i - L_f)/L_i = (\lambda_i - \lambda_f)/\lambda_i$

Therefore,

$$D_E = X (\lambda_i - \lambda_f)/\lambda_i$$

$$D_E (\lambda_i/\lambda_f)/X = (\lambda_i - \lambda_f)/\lambda_f$$

Because

$X = C_i t$

$\lambda_i = C_i/\nu_i$

Therefore,

$$D_E / (\lambda_f \nu_i)\ t = (\lambda_i - \lambda_f)/\lambda_f$$

Or

$$D/t = (\lambda\ \nu_1)\ (\lambda_1 - \lambda)/\lambda$$

$$D/t \propto (\lambda\ \nu_1)\ (l_{yy1} - l_{yy})/l_{yy}$$

Where D is the distance between the star and earth. λ_1 is the wavelength, ν_1 is the frequency and l_{yy1} is the Wu's Unit Length of the photon generated in the initial stage on the star. λ is the wavelength and l_{yy} is the Wu's Unit Length of the photon generated at the final stage on the present earth. t is the duration of the photon traveling from star to earth. $(\lambda_1-\lambda)/\lambda$ is the redshift and $(l_{yy1} - l_{yy})/l_{yy}$ is named "Wu's Spacetime Shrinkage Factor".

Also, the velocity of the reverse expansion V can be calculated as follows:

$V = (X/L' - X/L)L'/dt$

$V = X\ [-\ (L' - L)/L]/dt$

$$V = X\ (-dL/L)/dt$$

Because

$-dL/L = L\ dL^{-1}$

Also,

$L \propto l_{yy} \propto \lambda$

Therefore,

$$V = X\ (L\ dL^{-1})/dt$$

$$V = X\ (\lambda\ d\lambda^{-1})/dt$$

$$Vt = X \lambda_f (1/\lambda_f - 1/\lambda_i)$$

$$Vt = X (\lambda_f / \lambda_i) (\lambda_i - \lambda_f)/\lambda_f$$

Because

$\lambda_i = C_i/\nu_i$

$C_i t = X$

Therefore,

$$V = (\lambda_f / \nu_i) (\lambda_i - \lambda_f)/\lambda_f$$

Or

$$V = (\lambda/\nu_1) (\lambda_1 - \lambda)/\lambda$$

$$V = (\lambda/\nu_1) (l_{yy1} - l_{yy})/l_{yy}$$

Where V is the velocity of the reverse expansion. λ_1 is the wavelength, ν_1 is the frequency and l_{yy1} is the Wu's Unit Length of the photon generated in the initial stage on the star. λ is the wavelength and l_{yy} is the Wu's Unit Length of the photon generated at the final stage on the present earth. t is the duration of the photon traveling from star to earth. $(\lambda_1 - \lambda)/\lambda$ is the redshift and $(l_{yy1} - l_{yy})/l_{yy}$ is named "Wu's Spacetime Shrinkage Factor".

Because

$$D/t = (\lambda \nu_1) (\lambda_1 - \lambda)/\lambda$$

$$V = (\lambda/\nu_1) (\lambda_1 - \lambda)/\lambda$$

Therefore,

$$D/t = V \nu_1^2$$

Because

$\nu_1 = t_{yy1}^{-1} = \gamma^{-1} l_{yy1}^{-3/2}$

$D/t = V \gamma^{-2} l_{yy1}^{-3}$

$V = D/(t \gamma^{-2} l_{yy1}^{-3})$

Given

$$H_0 = \gamma^2 l_{yy1}^3/t$$

Then

$$V = H_0 D$$

For a star 5 billion years ago, Wu's Unit Length l_{yy1} converges to a constant and H_0 (Hubble Constant) becomes a time dependent constant.

As a result, Hubble's Law can also be derived from Wu's Spacetime Shrinkage Theory [67]. Because of this reason, instead of explained by the expansion of the universe due to the Acceleration Doppler Effect, Hubble's Law can also be interpreted by Wu's Spacetime Shrinkage Theory due to the aging of the universe. This is named "Wu's Spacetime Reverse Expansion Theory" [52].

12.5. The correlations of Wu Constant and Wu's Spacetime Constant to Hubble Constant

Furthermore, Wu Constant K and Wu's Spacetime Constant γ can be calculated and expressed by Hubble Constant H_0 as follows [73]:

Because

$H_0 = \gamma^2 l_{yy1}^3/t$

Therefore,

$$\gamma = (H_0 t)^{1/2} l_{yy1}^{-3/2}$$

Where γ is Wu's Spacetime Constant, H_0 is Hubble Constant, t is the time and l_{yy1} is the Wu's Unit Length of a star 5 billion light years away.

Also,

$\gamma = \pi (2K)^{-1/2}$

$K = \pi^2/2\gamma^2$

$$K = \pi^2 l_{yy1}{}^3/2H_0t$$

Where K is Wu Constant, γ is Wu's Spacetime Constant, H_0 is Hubble Constant, t is the time and l_{yy1} is the Wu's Unit Length of a star 5 billion light years away.

12.6. Wu's Spacetime Reverse Expansion Theory Versus Universe Expansion Theory

During Wu's Spacetime shrinkage process, the potential energy of Yangton and Yington circulating pairs can be converted to their kinetic energy with no need of external energy. Also, the distance between the star and earth remains unchanged at all time. There are no such things as that the star is undergoing acceleration and moving at a speed faster than the light speed. Because of these reasons, it is believed that Wu's Spacetime Reverse Expansion Theory based on Wu's Spacetime Shrinkage Theory is more realistic than Universe Expansion Theory in explanation of Cosmological Redshift and Hubble's Law.

12.7. Theories of Wu's Pairs, Photons and Corresponding Identical Objects and Events

Table 2 summarizes the theories of l_{yy}, t_{yy} and $V_{circulation}$ of Wu's Pairs; ν, C and λ of photons; and length, time, velocity and acceleration of Corresponding Identical Objects and Events with the influences of gravitational field and aging of the universe.

As a result, on the present earth (aged universe), because of the small Wu's Unit Length l_{yy} and Wu's Unit Time t_{yy}, photon has a smaller wave length λ and period T (larger frequency ν) than that coming from a star at far distance (young universe) which can cause redshift and universe expansion (more correctly spacetime shrinkage). Also, on a massive star (black hole), because of the large Wu's Unit Length l_{yy} and Wu's Unit Time t_{yy}, a corresponding identical object or event has a larger length and slower speed than that on earth.

Table 2 Theories of Wu's Pairs, Photons and Corresponding Identical Objects and Events

		Young Universe	Aged Universe	High Gravity	Low Gravity
Wu's Pairs					
l_{yy} (=2r)	l_{yy}	Large	Small	Large	Small
t_{yy} (=T)	$t_{yy} \propto l_{yy}^{3/2}$	Large	Small	Large	Small
V_{cir}(=2πr/T)	$V_{cir} \propto l_{yy}^{-1/2}$	Small	Large	Small	Large
Corresponding Identical Objects and Events					
L	$L \propto l_{yy}$	Large	Small	Large	Small
T	$T \propto l_{yy}^{3/2}$	Large	Small	Large	Small
V	$V \propto l_{yy}^{-1/2}$	Small	Large	Small	Large
A	$A \propto l_{yy}^{-2}$	Small	Large	Small	Large
Photons					
ν	$\nu = 1/t_{yy} \propto l_{yy}^{-3/2}$	Small	Large	Small	Large
C	$C \propto l_{yy}^{-1/2}$	Small	Large	Small	Large
λ	$\lambda = C/\nu \propto l_{yy}$	Large	Small	Large	Small

Chapter Thirteen

Einstein's Mistakes

Is Light Speed Always Constant?

Do Time, Length and Mass Change with Speed?

Since 1905, Einstein published his special relativity, general relativity, mass and energy conservation and field equations; the whole scientific world is confused with his theories on time dilations and interwoven spacetime. There are a few people who really understand Einstein's Theories even including Einstein himself. Despite the failure of proof of existence of aether by Michelson – Morley experiment, Einstein proposed that the light speed is constant in vacuum no matter of the light source and observer. He further claimed that time, momentum and mass can change with velocity (relativism) even that mass can become infinitive when travels at light speed (Really?). In addition, Einstein extended his special relativity theory to general relativity, in which he claimed that all physical properties change with acceleration even at constant speed (special relativity becomes a special case of general relativity where acceleration is zero). Furthermore, Einstein claimed that Spacetime is a solution of Einstein's Field Equations with acceleration as the curvature of Spacetime, instead of an outcome of gravitational field.

However, according to Yangton and Yington Theory, there are seven mistakes in Einstein's Theories including:

1. Light speed.

2. Special relativity and velocity time dilation.

3. Relativistic mass and Relativism.

4. General relativity and gravitational time dilation.

5. Spacetime.

6. Einstein's Field Equations.

7. Einstein's Law of Mass and Energy Conservation.

13.1. Light Speed

Einstein believed that light speed is constant and it doesn't change with the light source and observer. However, according to Yangton and Yington Theory, light speed is not constant. Instead, because of the photon Inertia Transformation, it is a vector summation of the Absolute Light Speed $\mathbf{C_S}$ (light speed observed at the light source) and Inertia speed $\mathbf{V_S}$ (light source speed observed by the observer).

$$\mathbf{C = C_S + V_S}$$

13.2. Special Relativity and Velocity Time Dilation

Einstein's Special Relativity is subject to two postulates: (1) All the laws of physics are the same in inertia systems, and (2) Light speed is always constant no matter the light source and observer. With these assumptions, Einstein claimed that the time (t) on a moving object is different from that observed on earth (t') which is called Velocity Time Dilation.

$$t' = 1/(1-V^2/C^2)^{1/2}\ t$$

However, because light speed is not constant and Velocity Time Dilation doesn't exist, therefore Einstein's Special Relativity and Velocity Time Dilation are fake theories.

13.3. Relativistic mass and Relativism

According to Special Relativity, Mass, Momentum, Energy and Length of a traveling object can also change with the speed of the object [69] when observed on earth. This is called Relativism.

$$M' = 1/(1-V^2/C^2)^{1/2} M$$

$$P' = 1/(1-V^2/C^2)^{1/2} MV$$

$$E^2 = M^2C^4 + P^2C^2$$

$$L' = (1-V^2/C^2)^{1/2} L$$

However, according to Yangton and Yington Theory, Mass is measured by the amount of Wu's Unit Mass (m_{yy}). Time is measured by the amount of Wu's Unit Time (t_{yy}) and Length is measured by the amount of Wu's Unit Length (l_{yy}). Also t_{yy} is related to l_{yy} by Wu's Spacetime Theory $t_{yy} = \gamma l_{yy}^{3/2}$, and l_{yy} is dependent on the gravitational field and aging of the universe. Therefore, Mass, Time and Length have nothing to do with the speed of the object. Again Einstein made a mistake by taking velocity as a principle factor in physics.

13.4. General Relativity and Gravitational Time Dilation

Einstein further extended his relativity theory to general relativity and Gravitational Time Dilation [47] using acceleration (curvature of spacetime) as a principle factor. He claimed that the special relativity is only a special case of general relativity where the acceleration is zero (constant velocity).

According to Yangton and Yington Theory, Mass, Time and Length are all dependent on the gravitational force and the aging of the universe. Because acceleration changes with all kind of forces, and gravitational force is only one of the Four Basic Forces, Einstein's general relativity is true only when acceleration is caused by the gravitational force [71]. In addition, Einstein missed totally the influence of Time and Length caused by the aging of the universe which results in the cosmological redshift, Hubble's Law and universe expansion (more correctly, spacetime shrinkage or spacetime reverse expansion), all because that he has absolutely no idea of Wu's Pairs and Yangton and Yington Theory in his time of 1910s.

13.5. Einstein's Spacetime

Einstein never really defined his spacetime. Einstein's Spacetime is relative and inextricably interwoven into what has become known as the space-time continuum. Unlike the Normal Spacetime and Wu's Spacetime, Einstein's Spacetime is not a reference system. It is a solution of Einstein's Field Equations with a four dimensional space-time continuum derived from a nonlinear geometry system to a Normal Spacetime System on earth. Reflecting the distribution of matter and energy, the derivative of the curvature of the space-time continuum represents the Amount of Normal Unit Acceleration in a Normal Spacetime System on earth.

In contrast, Wu's Spacetime [x, y, z, t](l_{yy}, t_{yy}) [3] is a special four dimensional system that is defined by the Wu's Unit Length l_{yy} (the diameter of Wu's Pairs) and Wu's Unit Time t_{yy} (the period of Wu's Pairs) at the reference point. Both Wu's Unit Length and Wu's Unit Time are dependent on the gravitational field and aging of the universe [70] [71] at the reference point. Also, they are correlated to each other by Wu's Spacetime Theory ($t_{yy} = \gamma l_{yy}^{3/2}$) [41].

Wu's Spacetime Field Equations observed on earth based on t_{yy0} and l_{yy0} have G and C_0^{-4} on the matter and energy side (right hand side) and the Amount of Normal Unit Acceleration "a_0" (the curvature) on the spacetime side (left hand side) of the equations, which is similar to Einstein's Field Equation.

$$a_0 = \delta\, \gamma^2\, C_0^{-4}\, G\, M/R^2$$

$$R_{\mu\nu} - \frac{1}{2} R\, g_{\mu\nu} + \Lambda\, g_{\mu\nu} = \frac{8\pi G}{c^4} T_{\mu\nu}$$

Therefore it is suggested that the curvature of Einstein's Spacetime (space-time continuum) in a Normal Spacetime on earth is in correspondance to the Amount of Normal Unit Acceleration in Wu's Spacetime on earth.

13.6. Einstein's Field Equations

Einstein's Field Equations have a solution with a four dimensional space-time continuum derived from a nonlinear geometry system to a Normal Spacetime System on earth. Reflecting the distribution of matter and energy, the derivative of the curvature of the space-time continuum represents the Amount of Normal Unit Acceleration in a Normal Spacetime System on earth.

In contrast, Wu's Field Equation represents the Amount of Normal Unit Acceleration reflecting the gravitational force and the distribution of matter in Wu's Spacetime on earth, where Wu's Unit Length l_{yy0} and Wu's Unit Time t_{yy0} are correlated to each other by $t_{yy0} = \gamma l_{yy0}^{3/2}$.

In other words, there is no gravitational force in Einstein's Spacetime. Acceleration is derived from the curvature of space-time continuum, which reflects the distribution of matter and energy in the universe. On the contrary, in Wu's Spacetime, matter does exist, as is the gravitational force, and the acceleration is indeed caused by the gravitational force.

Although Einstein understood that acceleration should relate to the absolute light speed (on earth) by C_0^{-4}, also the acceleration is the curvature of a space-time continuum in a Normal Spacetime System on earth, but he formulated his field equations and general relativity based on two wrong assumptions: (1) light speed is always constant, and (2) acceleration is the only principle factor in spacetime. In fact, spacetime is a function of Wu's Unit Time (t_{yy}) and Wu's Unit Length (l_{yy}) depending on the gravitational force and the aging of the universe at the reference point.

Because of the similarities that both G and C^{-4} are on the matter and energy side (right hand side) of the equations, Einstein's

Field Equations in a Normal Spacetime on earth can be considered as an equivalent form of Wu's Spacetime Field Equations based on Wu's Unit Length l_{yy0} and Wu's Unit Time t_{yy0} on earth [70] [71].

13.7. Einstein's Law of Mass and Energy Conservation

When a matter explodes, it becomes a bundle of free photons escaping into the space at a constant speed of 3×10^8 m/s. A massive energy in the magnitude of MC^2 is released. This theory is proposed by Einstein [26]. The theory predicts that matter and energy is interchangeable. Additionally a huge amount of energy can be released through the transformation (nuclear reaction).

Because photon is a free Wu's Pair traveling in space according to Yangton and Yington Theory, it is assumed that, during the explosion, a group of subatomic particles with mass M were first escaped into the space having kinetic energy ½ MC^2 at a light speed 3×10^8 m/s. And subsequently, Wu's Pairs were separated from the subatomic particles to form photons. Since all Wu's pairs become photons, the amount of Wu's Pairs remains unchanged during the explosion. Therefore, $E = MC^2$ has nothing to do with the transformation between mass and energy. In fact, it is an energy conversion from subatomic particle's structure energy (generated from string force and four basic forces) and kinetic energy to photon's kinetic energy $Mh\nu$ [68].

13.8. Einstein's Mistakes

Because Wu's Pairs are the building blocks of all matter, the time and length of an object or event can be measured by the local Wu's Unit Time and Wu's Unit Length depending on the local gravitational field and aging of the universe.

In addition, according to Principle of Correspondence [65], an object moving or event progressing under an equilibrium condition, the "Amounts of Wu's Unit Quantities" are always constant measured by the "Wu's Unit Quantities" no matter the gravitational field and aging of the universe. However, the total time and length can be different subject to the Wu's Unit Time and Wu's Unit Length at the same location.

Rather than the changes of time and length of a moving object or progressing event, Einstein believed that the Spacetime itself changes and twists when an object is moving under acceleration. As a result, Einstein derived his theories including Special Relativity, General Relativity, Spacetime, Field Equations and Mass and Energy Conservation, based on two wrong assumptions: (a) Light speed is always constant no matter the light source and observer, and (b) Acceleration is the principle factor of Spacetime.

In contrast, according to Yangton and Yington Theory, it is realized that (a) Light speed is not constant, instead, it is the vector summation of Absolute Light Speed C and Inertia Light Speed, and (b) Acceleration is not a principle factor, instead, gravitational field and aging of the universe are the principle factors of Wu's Spacetime. In other words, the time and length of a moving object or progressing event are a function of the local Wu's Unit Time (t_{yy}) and Wu's Unit Length (l_{yy}) depending on the local gravitational field and aging of the universe no matter of the acceleration [70][71].

Chapter Fourteen

The Beginning and End of the Universe

What Is a Singularity?

What Is a Black Hole?

How Will the Universe End?

14.1. What Is a Singularity?

Singularity [53] is the origin of energy and matter. Singularity is also the end of energy and matter. It is a tiny spot that could be found in the center of a Black Hole. During the Big Bang explosion, energy and matter were generated from a Singularity about 13.8 billion years ago.

Singularity is also a place that space and time can be generated and eliminated together in accompaniment with energy and matter. It is believed that a Singularity is a point gate entry to None, beyond the Singularity there is no space, time, energy or matter.

14.2. How the Universe Began?

In the beginning, there was only None. It was absolutely empty. There was no space, time, energy or matter. A Singularity was first created from None through Big Bang explosion. It is believed that space came from the Singularity before Wu's Pairs and matter were formed. If not, there was no place for Wu's Pairs to exist. Simultaneously energy came to exist together with space. Then, Wu's Pairs were formed from energy to build all the matter in the universe. At the same time, time was generated to allow the changes of the distribution of energy and the motion of matter in the universe.

14.3. What Is Black Hole?

A "Black Hole" (Fig. 32) [54] is a place in space where massive gravity is generated by the extremely large density caused by squeezing matter into a tiny space. This can happen in a dying star. Because no light can escape, Black Holes are not visible.

Due to the massive gravity, Wu's Pairs expand to form a hollow structure inside the Black Hole. At the center of the Black Hole, the balance between Force of Creation and centrifugal force is broken and the circulation of Wu's Pairs collapses. Subsequently, Yangton and Yington recombine and destroy each other, so that matter disappears and the universe returns back to an empty space. In other words, Black Holes are the grave yard of all matter. In the center of the Black Hole, there is a Singularity where everything goes back to None, and there is no space, time, energy or matter left.

Fig. 32 An image of Black Hole.

14.4. How will the Universe End?

At the end of the universe, it is proposed that Yangton and Yington will first recombine to destroy each other such that Energy of Creation and Energy of Circulation can be released to annihilate with space and everything will return back to None.

According to Yangton and Yington Theory, Wu's Pairs, a Yangton and Yington Circulating Pairs (the building blocks of all matter) could be ended in one of the following two ways:

1. Black Holes

In the Black Hole, the circulation of Yangton and Yington Pair is first destroyed by the massive gravitational force, followed by the recombination and destruction of Yangton and Yington Pairs, and then massive energy (Energy of Creation and Energy of Circulation) is released. Finally, energy annihilates with space and everything enters into a Singularity to become None.

2. Aging of the Universe

According to Wu's Spacetime Shrinkage Theory, after trillions of years, due to the shrinkage of Yangton and Yington Pairs, recombination and destruction between a Yangton and Yington Pair will occur. Finally, each Yangton and Yington Pair will form a tiny Singularity where Yangton and Yington Pair will convert to energy (Energy of Creation and Energy of Circulation), then annihilate with a tiny space to become None.

Chapter Fifteen

Nature of Space and Time

Can We Do Time Travel?

Does Space Have a Boundary?

15.1. Do Space and Time Have a Beginning?

Everything has a beginning. Therefore, both space and time must have beginnings. Since space came first, therefore before the beginning of time, space might have been either frozen or begun just one instant earlier than the beginning of time. If time is a secondary or induced element that reflects the changes of the distribution of energy and the motion of matter in the universe, then time should be created to accompany energy instead of space. In other words, space and energy must come before time.

15.2. Do Space and Time Have an End?

Everything has an end. Therefore, both space and time must have an end. However, after the end of time, space may be either frozen or last only one instant longer than the end of time. If time is a secondary or induced element, that reflects the changes of the distribution of energy and the motion of matter in the universe, then time should be ended in accompaniment with energy instead of space. In other words, time will end before space and energy.

15.3. Are Space and Time Continuous?

Space is a volume of vacant places. Without space, none of energy and matter can exist. If space is not continuous then what is in the gap between two spaces? The only possible answer is the None or a Singularity where it has no space, time, energy or matter. Also what is at the boundary of the universe? Again the answer is the None or a number of Singularities. There are many Black Holes in space and each Black Hole has a Singularity in the center. Space is continuous except those in the Singularities of Black Holes and at the boundary of the universe. A space totally isolated from the present space by Singularities and None cannot be realized in this universe. In other words, it is nothing but another universe.

Time is the changes of the distribution of energy and the motion of matter in the universe. If time is not continuous, then the distribution of energy and the motion of matter in the space are also discontinuous. In other words, energy and matter can jump from one place to other place at any time, which against the conservations of energy and matter. In addition, according to the 2nd law of thermodynamics, the total entropy of a system always increases, therefore time is not reversible. Furthermore, because of the circulation of Wu's Pairs, time is always continuous and non-interruptible. As a result, time is continuous and irreversible in accompaniment with the continuous space.

15.4. What Are the Finite Limits of Space and Time?

Since both space and time are continuous, inside the universe, there are no finite limits. However, in order to measure the finite space and time inside the universe, we need to use the smallest unit length and unit time that we can possibly find in the space in which we live. Because Wu's Pairs are the basic building blocks of all matter, it is obvious that Wu's Unit Length (l_{yy}) and Wu's Unit Time (t_{yy}) are the finite limits in measurements of space and time.

15.5. Does Space Have a Boundary?

Everything has an end, therefore the universe has a boundary. One can imagine space as a balloon and the balloon shell is the boundary. Outside the space there is None, with no space, time, energy or matter. Inside the space, there are billions of galaxies, stars and planets, and many Black Holes with Singularities at the centers, where space, time, energy and matter are all destroyed and become None. In other words, space disappears in the Singularities of the Black Holes. Since None can have a point entry (Singularity) to the universe, the whole balloon shell should be considered a network of Singularities. Therefore, the universe has a limited size with a boundary made of None or a network of Singularities.

15.6. Can Space Be Bent?

Space itself contains Nothing. Therefore, it is not bendable. However, a particle such as photon, X-ray and Graviton traveling linearly in space can be bent due to the influence of Spacetime caused by the gravitational field and aging of the universe.

15.7. How Is Time Related to Space?

There is no direct relationship between time and space. Time is a secondary or induced element from energy. It records the changes of distribution of energy and the motion of matter in space. The size l_{yy} (Wu's Unit Length) and the period t_{yy} (Wu's Unit Time) of Wu's Pairs are used to measure the dimension and time of the object, event and process in space. According to Wu's Spacetime Theory, t_{yy} is proportional to 3/2 power of l_{yy}.

15.8. Can Space Change Its Size?

Space was created in the beginning of the Big Bang explosion. It started to grow from a Singularity until all the energy was released. On the other hand, in each of the Black Holes, space, time, energy and matter recombine and destroy each other such that the total volume of the space can be reduced.

15.9. Is There a Wormhole?

A wormhole [55] (Fig. 33) or "Einstein-Rosen bridge" is a hypothetical topological feature that would fundamentally be a shortcut connecting two separate points in Spacetime. A wormhole may connect extremely long distances such as a billion light years or more, short distances such as a few feet, different universes, and different points in time. A wormhole is much like a tunnel with two ends, each at separate points in Spacetime.

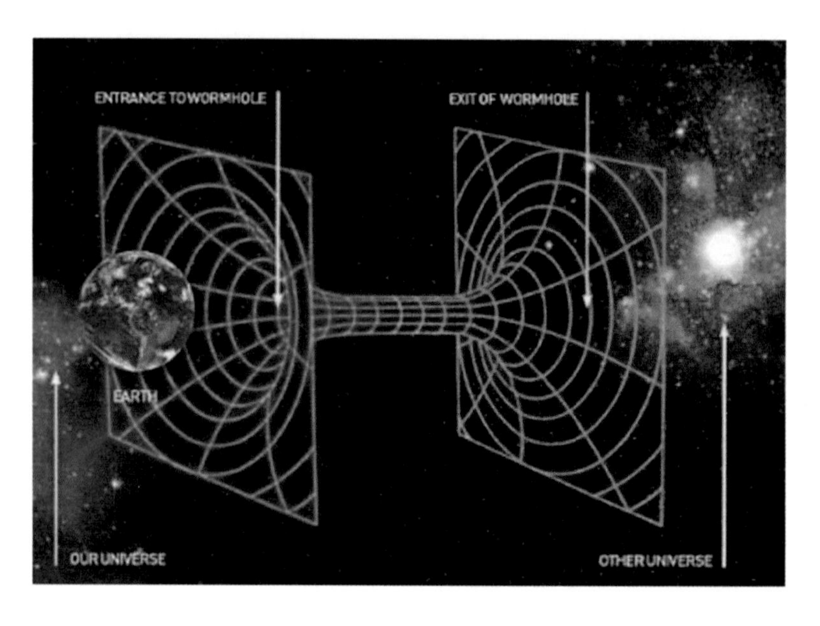

Fig. 33 An image of Wormhole.

For a simplified notion of a wormhole, space can be visualized as a two-dimensional (2D) surface. In this case, a wormhole would appear as a hole in that surface, lead into a 3D tube (the inside surface of a cylinder), then re-emerge at another location on the 2D surface with a hole similar to the entrance. An actual wormhole would be analogous to this, but with the spatial dimensions raised by one. For example, instead of circular holes on a 2D plane, the entry and exit points could be visualized as spheres in 3D space.

In Yangton and Yington Theory, Spacetime is a function of Wu's Unit Length (l_{yy}) and Wu's Unit Time (t_{yy}) of Wu's Pairs, which are related to each other by Wu's Spacetime Theory ($t_{yy} = \gamma l_{yy}^{3/2}$). Since Wu's Unit Length (l_{yy}) and Wu's Unit Time (t_{yy}) change with the gravitational field and aging of the universe, Spacetime is also a function of the gravitational field and aging of the universe.

However, the space in Yangton and Yington Theory is always a straight 3D Cartesian System. It is not curved or bent. In other words, Wormholes can't exist in the universe based on Yangton and Yington Theory.

15.10. Is There a Multiverse?

The Multiverse (Fig. 34) is the hypothetical set of possible universes, including the universe in which we live. Together, these universes comprise everything that exists: the entirety of space, time, matter and energy with the physical laws that describe them.

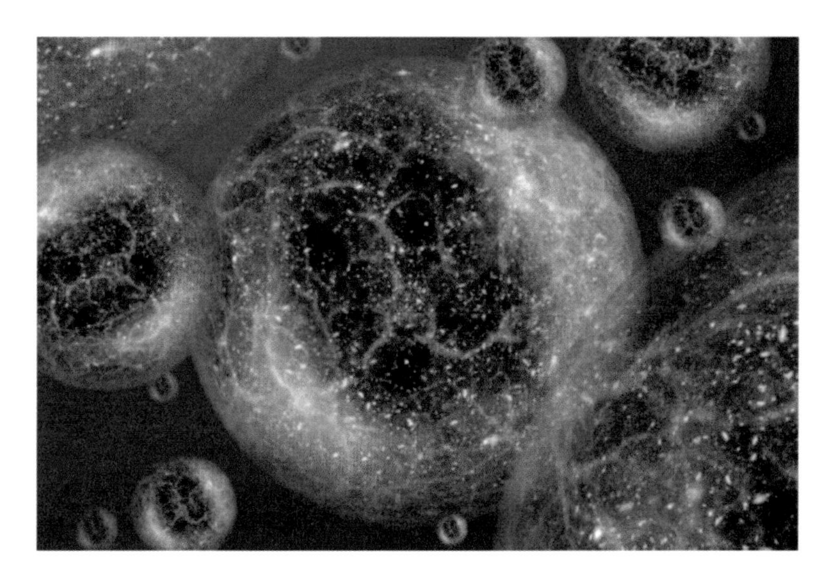

Fig. 34 An image of Multiverse.

In theory, Multiverse is possible and if one Big Bang explosion in None can create the universe in which we live, why can't another Big Bang creates another universe?

We can imagine the multiverse like bubbles with each universe having its own boundary made of None or a network of Singularities. Because two universes can only be connected to each other through None or a network of Singularities, therefore we can't never travel across the boundaries and enter into another universe.

15.11. Can We Do Time Travel?

Time reflects the changes of the distribution of energy and the motion of matter in the universe. If time is not continuous, then the distribution of energy and the motion of matter in the space are also discontinuous. In other words, energy and matter can jump from one place to other place at any time, which against the conservations of energy and matter. In addition, according to the 2nd law of thermodynamics, the total entropy of a system always increases, therefore time is not reversible. Furthermore, because of the circulation of Wu's Pairs, time is always continuous and non-interruptible. As a result, we cannot go back to the past nor jump into the future. Time travel is totally prohibited.

15.12. The End of the Universe

At the end of the universe, not only Yangton and Yington of Wu's Pairs will recombine to destroy each other either in the black holes or through the aging of the universe, but also space will annihilate with energy, such that everything will disappear and the whole universe will go back to None (no space, time, energy or matter). That would happen, at the final stage, through the last singularity, which will be the end of the universe.

Chapter Sixteen

Beyond My Universe

Is My Universe a Real Universe?

Is There a God?

16.1. Is My Universe a Real Universe?

Although "My Universe" is proposed based on a hypothetical "Yangton and Yington Theory" with "Wu's Pairs", a Yangton and Yington circulating pairs as the building blocks of all matter, some indirect evidences can be used to prove its reality, for examples:

1. Photon as a free Wu's Pair can be easily generated and emitted from a substance with thermal energy.

2. String structures made of Wu's Pairs by String Forces can be used to interpret String Theory.

3. Higgs Bosons can be considered as the String Force Carriers generated by Wu's Pairs and Higgs Field can be interpreted as the distribution of the String Force. This concurs that mass is the total amount of Wu's Pairs x Wu's Pair.

4. Gravitons, gravitational force and the propagation of gravitational force are explained by particle radiation and contact interaction.

5. De Broglie Matter Wave and Planck constant can be derived from a whirlpool model. Wave Particle Duality of photon and electron can be explained by Wu's Pairs and Yangton and Yington Theory.

6. Photon emission and Inertia Transformation can be used to derive Equation of Light Speed and explain Acceleration Doppler Effect and Event Horizon Theory.

7. Hubble's Law can be derived from Wu's Spacetime Shrinkage Theory without dark energy and Universe Expansion Theory can be interpreted by Wu's Shrinkage and Inverse expansion Theories.

8. Einstein's Gravitational Time Dilation, Spacetime, Field Equation and General Relativity are in accordance with Wu's Spacetime Shrinkage Theory, Spacetime Theory, Spacetime Field Equation and Principle of Correspondence.

9. Quantum Field Theory can be explained as a quantized field based on the particles of point structure and the distribution of particle radiation and contact interaction.

10. Quantum Gravity Theory can be interpreted as a quantized gravitational field based on the gravitons of string structure and the distribution of particle radiation and contact interaction.

11. Unified Field Theory can be explained as a quantized field generated by Four Basic Forces induced from Force of Creation between various subatomic particles with string structures made of Wu's Pairs.

12. Wu's Pairs and Yangton and Yington Theory can provide a particle model and physical picture for the interpretation of Quantum Field Theory, Quantum Gravity Theory and Unified Field Theory.

In principle, all theories that scientists have proposed to explain the phenomena of the universe are developed from their logical thinking, physical experiments, common experiences, and usually with some physical and mathematical models. However, the bottom line is how close those theories are to the real world and how good they are in explanation of the phenomena.

Based on the Five Principles of the Universe, it is proposed that Wu's Pairs, a Yangton and Yington circulation pair, with an inter-attractive Force of Creation are created from None where there is no space, time, energy or matter and what so ever. Wu's Pairs are the building blocks of all matter. With String Force generated between two adjacent Wu's Pairs induced from Force of Creation, String Structures (Elementary Subatomic Particles) are made in accordance with String Theory. Composite Subatomic Particles are then composed of various Elementary Subatomic Particles with Four Basic Forces including electromagnetic force, gravitational force, weak force and strong force in accordance with Unified Field Theory.

Gravitational force is generated between two gravitons with linear string structures. Electromagnetic force is created between electron and positron with spherical string structures. Both proton and neutron have ring structures. Weak force is formed between a pair of positron and neutron. Additionally, strong force is produced between two neutrons and a pair of neutron and proton.

The propagation of gravitational force like photon radiation is caused by the graviton radiation and contact interaction. Gravitational field reflects the distribution of the concentration of gravitons in space. Gravitational Wave is generated by the fluctuation of the graviton radiation from a pair of circulating massive stars or black holes.

Photon is a free Wu's Pair escaped from a substance through a two stage separation and ejection process. Upon Photon emission, Photon Inertia Transformation and Vision of Light, Equation of Light Speed can be derived as a vector summation of absolute light speed (3×10^8 m/s observed from light source) and inertia speed (the speed of light source observed by the observer). This opposes to Einstein's Special Relativity and Velocity Time Dilation based on the constant light speed.

Higgs Bosons can be considered as the String Force Carriers generated by Wu's Pairs and Higgs Field can be interpreted as the distribution of the String Force. This concurs that mass is the total amount of Wu's Pairs x Wu's Pair.

Furthermore, $E=MC^2$ is the transformation between matter's structure energy and photon's kinetic energy instead of the conversion between mass and energy.

According to the whirlpool model, where the momentum of a spinning particle is proportional to the mass and the spinning frequency of the particle, De Broglie Matter Wave, Planck constant and mass of Photon (Wu's Pair) can all be derived based on Yangton and Yington Theory.

Event Horizon can be explained by the competition between outward Absolute Light Speed and inward Photon Inertia Speed based on Photon Inertia Transformation. Redshift can be generated from Photon Inertia Transformation with Acceleration Doppler Effect. As a result, the expansion of the universe and Hubble's Law can be interpreted by the Acceleration Doppler Effect with Dark Energy. However, where the Dark Energy comes from remains a mystery.

When a photon emitted from a star, it maintains the same period and diameter as that of Wu's Pair in the parent star. In other words, photon carries the DNA of its parent star. According to Wu's Spacetime Theory, the period of Wu's Pair decreases with the diameter of Wu's Pair ($t_{yy} \propto l_{yy}^{3/2}$). Also, based on Wu's Spacetime Shrinkage Theory, the diameter of Wu's Pair can increase with the gravitational field and decrease with the aging of the universe. Because of these reasons, gravitational redshift and cosmological redshift can thus be generated. In fact, in addition to Acceleration Doppler Effect, the expansion of the universe and Hubble's Law can also be derived from Wu's Spacetime Shrinkage Theory and Wu's Reverse Expansion Theory without Dark Energy. Evenmore, Wu constant K and Wu's Spactime constant γ can be calculated and expressed by Hubble

Constant H_0. As a result, the Spacetime on earth is actually shrinking instead that the universe is expanding.

Since Wu's Pairs are the building blocks of all matter, in theory, a fundamental measurement can be applied based on Wu's Pair (unit mass), circulation period (unit time) and diameter (unit length) of Wu's Pairs.

Under equilibrium conditions, as an object or event takes place or moves to a different location, the dimensions of the object and the duration of the event will change with the gravitational field and aging of the universe, but not the structure of the object and the sequence of the event. These groups of objects and events are called "Corresponding Identical Objects" and "Corresponding Identical Events".

Each physical quantity of a corresponding identical object or event should have a constant amount of corresponding identical unit quantity, no matter the gravitational field and aging of the universe. This theory is named "Principle of Correspondence". As a result, all physical laws maintain unchanged to the corresponding identical objects and events observed in an inertia system. However, for an observation on earth, in addition to have larger length and time, an object or event on a massive star (black hole) with large gravitational field has smaller velocity and acceleration comparing to that of the corresponding identical object or event on earth with less gravitational field. This result agrees very well with general relativity.

According to Principle of Correspondence, both deflection of light and Perihelion Precession of Mercury are resulted from the decreasing speeds of photon and Mercury, while passing through a massive star (sun), due to the large Wu's Unit Length ($V \propto l_{yy}^{-1/2}$) caused by an extremely large gravitational force.

Finally, Wu's Spacetime and Wu's Field Equations are derived based on Wu's Unit Time and Wu's Unit Length obeying Wu's Spacetime Theory and depending on the gravitational field and aging of the universe. In comparison, Einstein's Spacetime is a solution of Einstein's Field Equations with space-time continuum derived from a nonlinear geometry system.

To be more specifically, Wu's Pairs and Yangton and Yington Theory can be used successfully in explanation and derivation of the following major physical phenomena and theories:

1. Five Principles of the Universe [59].

2. Wu's Pairs and Force of Creation [1].

3. Photons as free Wu's Pairs [1].

4. String Theory, String Force and String Structures [2].

5. Subatomic Particle and Dark Matter based on String Structures and Four Basic Forces [2].

6. Four Basic Forces and Unified Field Theory based on Force of Creation [2] [60].

7. Antimatter and Baryogenesis based on Wu's Pairs [2].

8. Graviton Structures and Gravitational Force based on Wu's Pairs and Force of Creation [2].

9. Graviton Radiation and Contact Interaction and Newton's Law of Universal Gravitation [17].

10. Gravitational Field interpreted by Distribution of Graviton Concentration [17].

11. Quantum Field Theory interpreted by Graviton Radiation and Contact Interaction [17].

12. Gravitational Waves and Graviton Radiation [17].

13. Higgs Bosons considered as the String Force Carriers generated by Wu's Pairs; Higgs Field explained as the

distribution of the String Force and Wu's Pairs; Mass interpreted as the amount of Wu's Pairs x Wu's Pair [75].

14. Masses of Wu's Pair, Photon and Subatomic Particles [28].

15. $E = MC^2$ as the energy transformation between Wu's Pairs and Photons [26].

16. Photon Emission and Absolute Light Speed [28].

17. Photon Inertia Transformation and Inertia Light Speed [28].

18. Black Body Radiation and Wu's Pairs [28].

19. Vision of Object, Vision of Light and Theory of Vision [35].

20. Equation of Light Speed [35].

21. Mistakes of Special Relativity and Velocity Time Dilation [34].

22. Mistakes of General Relativity and Gravitational Time Dilation [71].

23. Acceleration Doppler Effect and Doppler Redshift [29].

24. Hubble's Law and Acceleration Doppler Effect [29].

25. Wu's Unit Mass, Wu's Unit Time and Wu's Unit Length [65].

26. Principle of Correspondence [65].

27. Deflection of Light [Annex 31] and Perihelion Precession of Mercury [Annex 32] Interpreted by Yangton and Yington Theory [80].

28. Wu's Spacetime [41].

29. Wu's Spacetime Theory [41].

30. Wu's Spacetime Field Equations [57].

31. Photon and Wu's Spacetime [41].
32. Cosmological Redshift and Wu's Spacetime Shrinkage Theory based on Aging of the Universe [41].
33. Gravitational Redshift and Wu's Spacetime Shrinkage Theory based on Gravitational Field [41].
34. Hubble's Law and Wu's Spacetime Shrinkage Theory [52].
35. Wu's Spacetime Reverse Expansion Theory [41].
36. The correlations of Wu Constant and Wu's Spacetime Constant to Hubble Constant [73].
37. Derivations of Planck Constant, De Broglie Wave and Mass of Photon (Wu's Pair) Based on Yangton and Yington Theory [74].
38. Destruction of Wu's Pairs by aging of the universe [41].
39. Destruction of Wu's Pairs in Black Hole by gravitational force [41].
40. Wu's Spacetime versus Einstein's Spacetime [70].
41. Wu's Spacetime Field Equations versus Einstein's Field Equations [70][81].
42. The hollow structure of Black Hole based on Wu's Spacetime Theory [70].
43. Event Horizon and Black Hole interpreted by Photon Inertia Transformation [76] and Wu's Spacetime Field Equations [70].
44. Einstein's Seven Mistakes resulted from his two wrong assumptions: (1) Light Speed is always constant, and (2) Acceleration is the principle factor of the universe [71].

45. Corresponding Identical Objects and Events in large Gravitational Field Observed on Earth [72].

A road map of the systematic derivations and correlations between the major physical phenomena and Yangton and Yington Theory can be shown in Fig. G [66]:

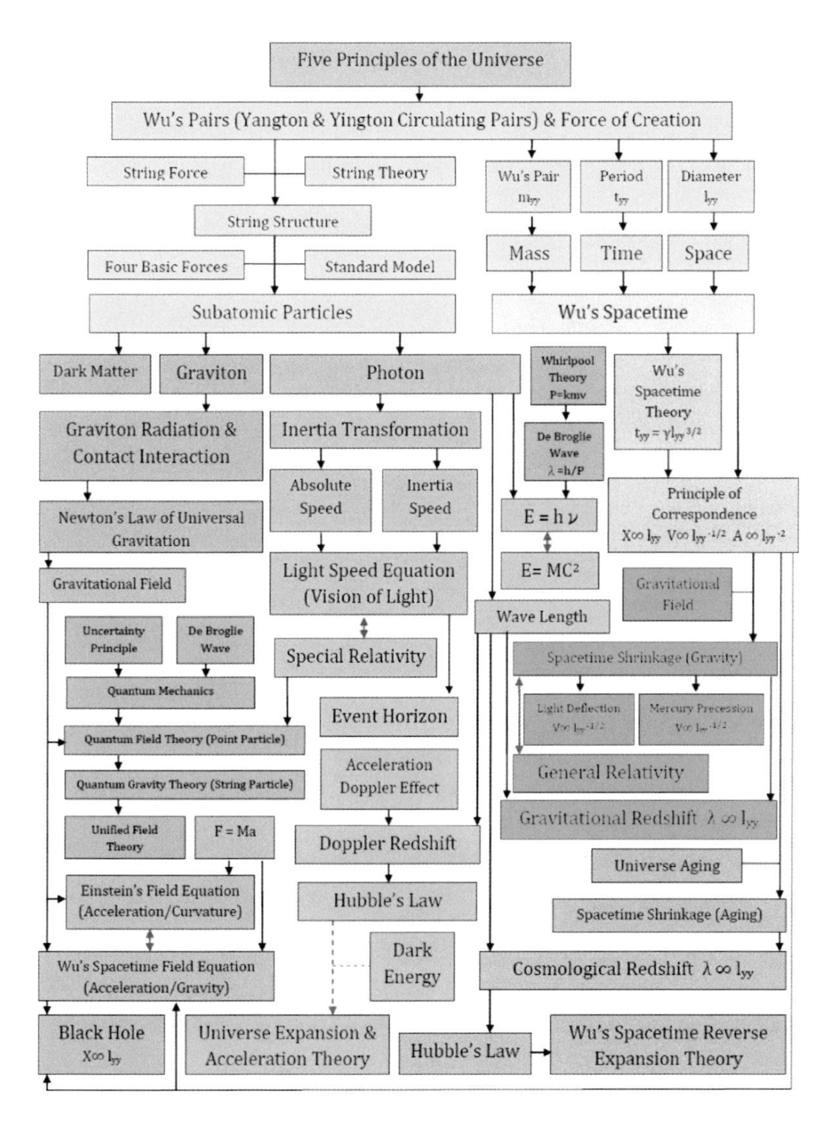

Fig. G A road map of systematic derivations and correlations between major physical phenomena and Yangton and YingtonTheory.

In addition to be an excellent model revealing the real universe, Wu's Pairs and Yangton and Yington Theory can be considered as a binary universe. Just like the binary system to the decimal system in mathematics, many theories and principles developed in the binary universe can be used effectively explaining the real universe.

However for those questions beyond the physical world such as how the space, time, energy and matter are created, even there are no good answers, I would like to share some of my thoughts with readers as follows:

16.2. Creation of Space and Energy

Among the four basic elements of the universe: space, time, energy and matter, it is believed that space and energy are two primary elements. Time and matter are two secondary ones or induced elements. Matter is the distribution of energy and time is the change of the distribution of energy and the motion of matter.

During Big Bang explosion, space and energy were first created together in the Singularity from None. The process should be reversible such that space and energy can recombine and destroy each other to ensure that everything will return back to None. This is called "Annihilation of Space and Energy".

$$\text{None} \leftrightarrow \text{Space} + \text{Energy}$$

Wu's Pairs were then formed from energy and subsequently all other matter were formed from Wu's Pairs. Thus the universe was born.

16.3. Creation of Matter

It is proposed that a number of super fine Yangton and Yington Antimatter particle pairs with an inter-attractive Force of Creation were generated in the Singularity by absorbing energy from the Big Bang explosion. Also, because of the enforcement of the inter-attractive Force of Creation, Yangton and Yington particles can recombine and destroy each other such that Something can go back to Nothing.

By absorbing external energy from Big Bang explosion, the temporary Yangton and Yington Pairs can become a permanent Wu's Pairs with a circulation balanced between the centrifugal force and the inter-attractive Force of Creation.

Instead of a solid particle, Yangton and Yington can also be considered as two tiny energy whirlpools (energy particles) with opposite spin up (Yangton) and spin down (Yington) directions.

Once Wu's Pairs were formed, all subatomic particles such as photons, quarks, electrons, neutrons and protons, with string force and four basic forces including gravitational force, electromagnetic force, weak force and strong force can be generated from Wu's Pairs and inter-attractive Force of Creation. Simple atoms were then produced and finally stars and galaxies were formed and the entire universe was born.

16.4. Creation of Time

Time is a secondary element of the universe. It reflects the change of the distribution of energy and the motion of matter. Without energy there would be no time. Therefore, time is formed in accompaniment with energy.

16.5. What Happed in the Big Bang Explosion?

The Big Bang Theory (Fig. 35) is the prevailing cosmological model for the universe from the earliest known periods through its subsequent large-scale evolution. The model accounts for the fact that the universe expanded from a very high density and high temperature state, and offers a comprehensive explanation for a broad range of phenomena, including the abundance of light elements, the cosmic microwave background, large scale structure and Hubble's Law. If the known laws of physics are extrapolated to the highest density regime, the result is a Singularity that is typically associated with the Big Bang explosion 13.8 billion years ago. After the initial expansion, the universe cooled sufficiently to allow the formation of subatomic particles, and later simple atoms. Giant clouds of these primordial elements later coalesced through gravity in halos of Dark Matter, eventually forming the stars and galaxies visible today.

According to Yangton and Yington Theory, it is assumed that not only energy and matter, but also space and time were all created at the same time from a Singularity in the beginning of the Big Bang explosion. Wu's Pairs, a Yangton and Yington circulating pairs with an inter-attractive Force of Creation, were first formed by two tiny energy whirlpools (energy particles) with opposite spin directions. Subsequently, subatomic particles were formed by Wu's Pairs with string force and four basic forces. Furthermore, once the first Yangton and Yington Pair were formed, time was created to reflect the changes of the distribution of energy and the motion of matter.

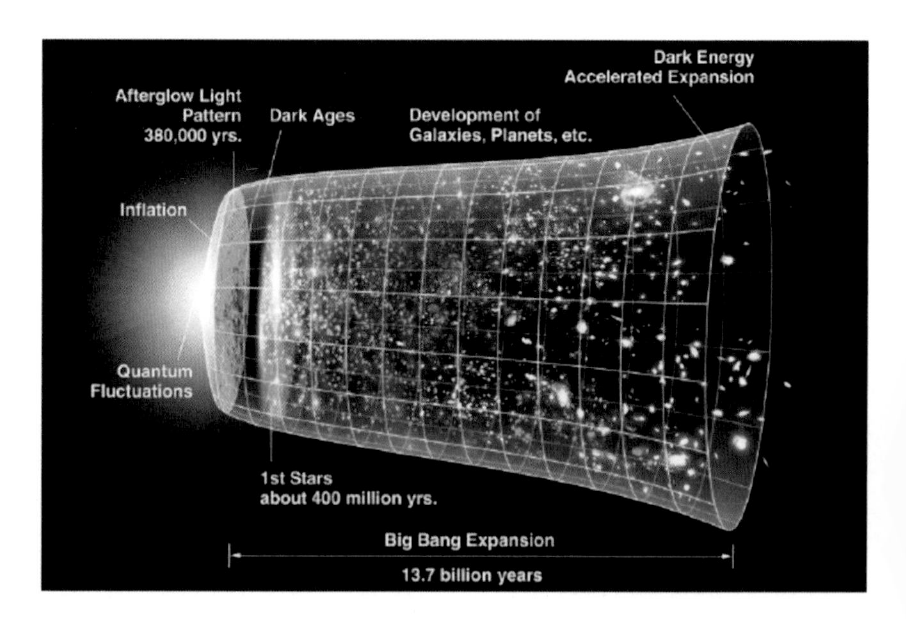

Fig. 35 Big Bang and the expansion of the universe.

16.6. Is There a God?

I don't have an exact answer for this question. It is my opinion that as long as there are major unanswered questions about the universe such as how the space and energy formed and where the energy in the Big Bang came from, it is very possible that there is an Intelligence (God?) at somewhere (None?) who created our universe including space, time, energy and matter. Otherwise, there will be no answer except that may be "God is the None" and "None is the God"?

16.7. Future Study

Wu's Pairs and Yangton and Yington Theory seem ideal in explanation of the structures and properties of most subatomic particles and major phenomena in the universe. However, because that Wu's Pairs are too small to study even with the present state-of-the-art scientific instruments, there is no proof of the existence of Wu's Pairs and if they can be used to form the subatomic particles. For future research, quantum mechanics and quantum field theory will be used to study the interactions between Wu's Pairs and the formation of subatomic particles by Wu's Pairs. Also, in correspondence with String Theory, quantum gravity theory will be interpreted based on the gravitons with string structures made of Wu's Pairs and the theory of particle radiation and contact interaction. In addition, similar to Einstein's Field Equations, a mathematical model based on nonlinear geometry will be developed to correlate the curvature of the space-time continuum to the Amount of Normal Unit Acceleration in Wu's Spacetime System. Both the correlations of Wu's Unit Length l_{yy} to gravitational field and aging of the universe will be studied in compliance with redshift and Hubble's Law. Also, the correlation between light speed and graviton concentration will be analyzed. Furthermore, the spins of Yangton and Yington and their correlations to that of subatomic particles will be studied. Hopefully with all these efforts, Wu's Pairs and Yangton and Yington Theory can be better understood and will be accepted by the scientific world in the near future.

References

[1] Edward T. H. Wu, "Yangton and Yington – A Hypothetical Theory of Everything", Science Journal of Physics, Volume 2015, Article ID sjp-242, 6 Pages, 2015, doi: 10.7237/sjp/242.

[2] Edward T. H. Wu. "Subatomic Particle Structures and Unified Field Theory Based on Yangton and Yington Hypothetical Theory". American Journal of Modern Physics. Vol. 4, No. 4, 2015, pp. 165-171. doi: 10.11648/j.ajmp. 20150404.13.

[3] "Big-bang model". Encyclopedia Britannica. Retrieved 11February 2015.

[4] Polchinski, Joseph (1998). String Theory, Cambridge University Press ISBN 0521672295.

[5] Highfield, Roger. "Large Hadron Collider: Thirteen ways to change the world". The Daily Telegraph. London. Retrieved 2008-10-10.

[6] "Subatomic Particle" Encyclopedia Britannica. Retrieved 2008-06-29.

[7] http://en.wikipedia.org/wiki/Higgs_boson.

[8] https://en.wikipedia.org/wiki/Neutron.

[9] https://en.wikipedia.org/wiki/Proton.

[10] J. Christman (2001). "The Weak Interaction" (PDF). Physnet. Michigan State University.

[11] Chapter 4 NUCLEAR PROCESSES, THE STRONG FORCE, M. Ragheb 1/27/2012, University of Illinois.

[12] https://en.wikipedia.org/wiki/Antimatter.

[13] Dark Matter. CERN. Retrieved on 17 November 2014.

[14] Purcell, E. (2011). Electricity and Magnetism (2nd ed.). Cambridge University Press. pp. 173 – 4. ISBN 1107013607.

[15] Beyond Art: A Third Culture page 199. Compare Uniform field theory.

[16] Chandrasekhar, Subrahmanyan (2003). Newton's Principia for the common reader.Oxford:Oxford University Press. pp. 1 – 2.

[17] Edward T. H. Wu. "Gravitational Waves, Newton's Law of Universal Gravitation and Coulomb's Law of Electrical Forces Interpreted by Particle Radiation and Interaction Theory Based on Yangton & Yington Theory". American Journal of Modern Physics. Vol. 5, No. 2, 2016, pp. 20-24. doi:10.11648/j.ajmp.20160502.11.

[18] https://en.wikipedia.org/wiki/Inverse-square_law.

[19] Coulomb (1785a) "Premier memoire sur l' electricite et le magnetisme,"Histoire de l' Academie Royale des Sciences, pages 569-577.

[20] https://en.wikipedia.org/wiki/Gravitational_wave.

[21] Einstein, A (June 1916). "Naherungsweise Integration der Feldgleichungen der Gravitation". Sitzungsberichte der Koniglich Preussischen Akademie der Wissenschaften Berlin. part 1: 688 – 696.

[22] B.P. Abbott et al. (LIGO Scientific Collaboration and Virgo Collaboration) (2016). "Observation of Gravitational Waves from a Binary Black Hole Merger". Physical Review Letters 116 (6). Bibcode: 2016 PhRvL.116f1102A. doi: 10.1103/PhysRevLett.116.061102.

[23] https://en.wikipedia.org/wiki/Momentum.

[24] Andrew Motte translation of Newton's Principia (1687) Axioms or Laws of Motion.

[25] Richard Feynman (1970). The Feynman Lectures on Physics Vol I. Addison Wesley Longman. ISBN 978-0-201-02115-8.

[26] Einstein, A. (1905), "Ist die Trägheit eines Körpers von seinem Energie inhalt abhängig?", Annalen der Physik, 18: 639–643, Bibcode: 1905
AnP...323..639E, doi:10.1002/andp.19053231314. See also the English translation.

[27] Reitz, John; Milford, Frederick; Christy, Robert (1992). Foundations of Electromagnetic Theory (4th ed.). Addison Wesley. ISBN 0-201- 52624-7.

[28] Edward T. H. Wu. "Mass, Momentum, Force and Energy of Photon and Subatomic Particles, and Mechanism of Constant Light Speed Based on Yangton & Yington Theory". American Journal of Modern Physics. Vol. 5, No. 4, 2016, pp. 45-50. doi: 10.11648/j.ajmp.20160504.11.

[29] Edward T. H. Wu, Redshift Caused by Acceleration Doppler Effect and Hubble's Law Based on Wu's Spacetime Accelerating Shrinkage Theory, American Journal of Modern Physics. Vol. 6, No. 1, 2017, pp. 10-15. doi: 10.11648/j.ajmp.20170601.12.

[30] Max Born and Emil Wolf, 1999, Principles of Optics, Cambridge University Press, Cambridge.

[31] "Shoaling, Refraction, and Diffraction of Waves". University of Delaware Center for Applied Coastal Research. Retrieved 2009-07 -2317. Hecht, Eugene. Optics, 2nd ed., Addison Wesley (1990) ISBN 0-201-11609-X. Chapter 8.

[32] Hecht, Eugene. Optics, 2nd ed., Addison Wesley (1990) ISBN 0-201-11609-X. Chapter 8.

[33] Peter Theodore Landsberg (1990). "Chapter 13: Bosons: black-body radiation". Thermodynamics and statistical

mechanics (Reprint of Oxford University Press 1978 ed.). Courier Dover Publications. pp. 208 ff. ISBN 0-486-66493-7.

[34] Edward T. H. Wu. "Light Speed in Vacuum is not a Constant and Time Doesn't Change with Velocity - Discrepancies Between Relativities and Yangton & Yington Theory". American Journal of Modern Physics. Vol. 4, No. 6, 2015, pp. 367-373. doi: 10.11648/j.ajmp.20150406.12.

[35] Edward T. H. Wu. "Vision of Object, Vision of Light, Photon Inertia Transformation and Their Effects on Light Speed and Special Relativity." IOSR Journal of Applied Physics (IOSR-JAP), vol. 9, no. 5, 2017, pp. 49–54.

[36] Michelson, Albert A.; Morley, Edward W. (1887). "On the Relative Motion of the Earth and the Luminiferous Ether". American Journal of Science.34: 333 345. doi:10.2475/ajs.s3-34.203.333.

[37] Einstein, Albert (1936). "Lens-like Action of a Star by the Deviation of Light in the Gravitational Field". Science. 84 (2188): 506–7. Bibcode: 1936Sci .84.506E. doi: 10.1126/science. 84.2188.506. JSTOR 1663250. PMID 17769014.

[38] Alec Eden The search for Christian Doppler, Springer-Verlag, Wien 1992. Contains a facsimile edition with an English translation.

[39] Einstein A. (1916), Relativity: The Special and General Theory (Translation 1920), New York: H. Holt and Company.

[40] Kuhn, Karl F.; Theo Koupelis (2004). In Quest of the Universe. Jones & Bartlett Publishers. pp. 122-3. ISBN 0-7637-0810-0.

[41] Edward T. H. Wu. "Time, Space, Gravity and Spacetime Based on Yangton & Yington Theory, and Spacetime Shrinkage Versus Universe Expansion". American Journal of Modern Physics. Vol. 5, No. 4, 2016, pp. 58-64. doi: 10.11648/j.ajmp.20160504.13.

[42] Peebles, P. J. E. and Ratra, Bharat (2003). "The cosmological constant and dark energy". Reviews of Modern Physics 75 (2): 559-606. arXiv: astro-ph/0207347. Bibcode: 2003 RvMP.75.559 P. doi: 10.1103/RevModPhys.75.559.

[43] https://en.wikipedia.org/wiki/Length_contraction.

[44] Einstein, Albert (1920). "On the Idea of Time in Physics". Relativity: The Special and General Theory. Henri Holt. ISBN 1-58734-092-5. And also in sections 9 – 12.

[45] https://en.wikipedia.org/wiki/Twin_paradox.

[46] https://en.wikipedia.org/wiki/General_relativity.

[47] A. Einstein, "Uber das Relativitatsprinzip und die aus demselben gezogenen Folgerungen", Jahrbuch der Radioaktivitat und Elektronik 4, 411 – 462 (1907); also in H M Schwartz, "Einstein's comprehensive 1907 essay on relativity, part I", American Journal of Physics, vol.45,no.6 (1977) pp.512 – 517; Part II, vol.45 no.9 (1977), pp.811 – 817; Part III, vol.45 no.10 (1977), pp.899 – 902.

[48] Catoni, F.; et al. (2008). Mathematics of Minkowski Space. Frontiers in Mathematics. Basel: Birkhauser Verlag. doi: 10.1007/978-3-7643-8614-6. ISBN 978-3-7643-8613-9. ISSN1660-8046.

[49] Slipher, Vesto (1915). "Spectrographic Observations of Nebulae".Popular Astronomy 23:21-24.Bibcode:1915 PA.23.21 S.

[50] Hubble, E. (1929). "A relation between distance and radial velocity among extra-galactic nebulae". Proceedings of the National Academy of Sciences.15(3): 168 - 73. Bibcode:1929PNAS...15..168H. doi:10.1073/pnas.15.3.168. PMC 522427. PMID 16577160.

[51] Frieman, Joshau A.; Turner, Michael S.; Huterer, Dragan. "Dark Energy and the Accelerating Universe" (PDF). Annu. Rev. Astron. Astrophys. arXiv: 0803.0982 v 1. Retrieved April 1, 2016.

[52] Edward T. H. Wu "Hubble's Law Interpreted by Acceleration Doppler Effect and Wu's Spacetime Reverse Expansion Theory." IOSR Journal of Applied Physics (IOSR-JAP), vol. 10, no. 1, 2018, pp. 58-62.

[53] Claes Uggla (2006). "Spacetime Singularities". Einstein Online. **2** (1002).

[54] Wald, R. M. (1997). "Gravitational Collapse and Cosmic Censorship". In Iyer, B. R.; Bhawal, B. (eds.). Black Holes, Gravitational Radiation and the Universe. Springer. pp. 69–86.

[55] Morris, Michael; Thorne, Kip; Yurtsever, Ulvi (1988). "Wormholes, Time Machines, and the Weak Energy Condition" (PDF). Physical Review Letters. 61(13): 1446–1449. Bibcode:1988PhRvL..61.1446M. doi:10.1103/PhysRevLett.61.1446. PMID 10038800.

[56] Carroll, Sean (2004). Spacetime and Geometry – An Introduction to General Relativity. pp. 151–159. ISBN 0-8053-8732-3.

[57] Edward T. H. Wu "Wu's Spacetime Field Equation Based On Yangton And Yington Theory." IOSR Journal of Applied Physics (IOSR-JAP), vol. 10, no. 2, 2018, pp. 13-21.

[58]EthanSiegelhttps://www.forbes.com/sites/startswithabang/2018/06/29/ surprise-the-hubble-constant-changes-over-time/#5e6406409c9a.

[59] Edward T. H. Wu" Five Principles of the Universe and the Correlations of Wu's Pairs and Force of Creation to String Theory and Unified Field Theory." IOSR Journal of Applied Physics (IOSR-JAP), vol. 10, no. 4, 2018, pp. 17-21.

[60] Edward T. H. Wu "Standard Model and Quantum Field Theory versus Wu's Pairs and Yangton and Yington Theory." IOSR Journal of Applied Physics (IOSR-JAP), vol. 10, no. 4, 2018, pp. 50-56.

[61] Cao, Tian Yu & Schweber, Silvan S. "The Conceptual Foundations and the Philosophical Aspects of Renormalization Theory", Synthese, 97(1) (1993), 33–108.

[62] Rovelli, C. (2004). Quantum Gravity. Cambridge Monographs on Mathematical Physics. p. 71. ISBN 978-0-521-83733-0.

[63] "The Standard Model – Particle decays and annihilations". The Particle Adventure: The Fundamentals of Matter and Force. Lawrence Berkeley National Laboratory. Retrieved 17 October2011.

[64] Zee, Anthony (2010). Quantum Field Theory in a Nutshell (2nd ed.). Princeton University Press. ISBN 978-0691140346.

[65] Edward T. H. Wu "Mass, Time, Length, Vision of Object and Principle of Correspondence Based on Yangton and Yington Theory" IOSR Journal of Applied Physics (IOSR-JAP), vol. 10, no. 5, 2018, pp. 80-84.

[66] Edward T. H. Wu "A Summary of Yangton and Yington Theory and Their Interpretations on Subatomic Particles, Gravitation and Cosmology" IOSR Journal of Applied Physics (IOSR-JAP), vol. 10, no. 5, 2018, pp. 45-50.

[67] Edward T. H. Wu "Hubble's Law Derived from Wu's Spacetime Shrinkage Theory and Wu's Spacetime Reverse Expansion Theory versus Universe Expansion Theory." IOSR Journal of Applied Physics (IOSR-JAP), vol. 11, no. 1, 2019, pp. 03-07.

[68] Edward T. H. Wu. "Einstein's $E = MC^2$ as Energy Conversion Instead of Mass and Energy Conservation and Energy and Space

Annihilation Based on Yangton and Yington Theory." IOSR Journal of Applied Physics (IOSR-JAP), vol. 11, no. 2, 2019, pp. 57-61.

[69] Roche, J (2005). "What is mass?" (PDF). European Journal of Physics. 26 (2):225. Bibcode:2005EJPh...26..225R. doi:10.1088/0143-0807/26/2/002.

[70] Edward T. H. Wu. "Einstein's Spacetime and Einstein's Field Equations Versus Wu's Spacetime and Wu's Spacetime Field Equations." IOSR Journal of Applied Physics (IOSR-JAP), vol. 11, no. 2, 2019, pp. 13-18.

[71] Edward T. H. Wu. "Einstein's Seven Mistakes." IOSR Journal of Applied Physics (IOSR-JAP), vol. 11, no. 3, 2019, pp. 15-17.

[72] Edward T. H. Wu. "General Relativity versus Yangton and Yington Theory – Corresponding Identical Objects and Events in Large Gravitational Field Observed on Earth." IOSR Journal of Applied Physics (IOSR-JAP), vol. 11, no. 3, 2019, pp. 41-45.

[73] Edward T. H. Wu "The correlations of Wu Constant and Wu's Spacetime Constant to Hubble Constant" IOSR Journal of Applied Physics (IOSR-JAP) , vol. 11, no. 4, 2019, pp. 01-06.

[74] Edward T. H. Wu."Derivations of Planck Constant and De Broglie Matter Waves from Yangton and Yington Theory." IOSR Journal of Applied Physics (IOSR-JAP), vol. 11, no. 5, 2019, pp. 68-72.

[75] Edward T. H. Wu "Higgs Boson and Graviton Interpreted by String Force and String Structures Based on Wu's Pairs and Yangton and Yington Theory" IOSR Journal of Applied Physics (IOSR-JAP) , vol. 11, no. 6, 2019, pp. 51-55.

[76] Edward T. H. Wu". Event Horizon and Black Hole Interpreted by Photon Inertia Transformation and Yangton and Yington Theory." IOSR Journal of Applied Physics (IOSR-JAP), vol. 11, no. 6, 2019, pp. 58-61.

[77] Heisenberg, W. (1927), "Über den anschaulichen Inhalt der quantentheoretischen Kinematik und Mechanik", Zeitschrift für Physik (in German), 43 (3–4): 172–198.

[78] Dyson, F. W.; Eddington, A. S.; Davidson C. (1920). "A determination of the deflection of light by the Sun's gravitational field, from observations made at the total eclipse of 29 May 1919". Philosophical Transactions of the Royal Society 220A(571–581): 291–333. Bibcode: 1920RSPTA.220..291D. doi: 10.1098/rsta.1920.0009.

[79] https://en.wikipedia.org/wiki/Tests_of_general_relativity# Perihelion_precession_of_Mercury.

[80]Edward T. H. Wu. "Perihelion Precession of Mercury and Deflection of Light Interpreted by Yangton and Yington Theory." IOSR Journal of Applied Physics (IOSR-JAP), 12(1), 2020, pp. 20-26.

[81] Edward T. H. Wu. "A Summary of Wu's Spacetime Field Equation and Its Comparison to Einstein's Field Equation" IOSR Journal of Applied Physics (IOSR-JAP), 12(1), 2020, pp. 09-19.

Annex 1

Science Journal of Physics
ISSN: 2276-6367
http://www.sjpub.org

Published By
Science Journal Publication
International Open Access Publisher

Research Article Volume 2015, Article ID sjp-242, 6 Pages, 2015, doi: 10.7237/sjp/242

Yangton and Yington – A Hypothetical Theory of Everything

Edward T. H. Wu

Solar Buster Corporation
USA (818) 991-7437

E-mail: edwardthwu@yahoo.com

Accepted on March 17, 2015

Abstract: A hypothetical theory of a Yangton and Yington circulating pair with an inter-attractive Force of Creation is proposed to explain the creation of the Universe and the formation of light and Matter. This theory can combine Particle Physics and Quantum Mechanics, unify Four Forces, imply Graviton Particles, implement String Theory, define Black Hole as well as provide proof to Einstein's Theory $E = MC^2$. It is also assumed that Dark Matter is made of a multiple of Yangton and Yington circulating pairs.

Keywords: Yangton, Yington, Photon, Subatomic Particles, Particle Physics, Quantum Mechanics, Four Forces, Graviton, String Theory, Black Hole, Einstein's Theory and Dark Matter.

Annex 2

American Journal of Modern Physics
2015; 4(4): 189-195
Published online June 29, 2015 (http://www.sciencepublishinggroup.com/j/ajmp)
doi: 10.11648/j.ajmp.20150404.15
ISSN: 2326-8867 (Print); ISSN: 2326-8891 (Online)

Subatomic Particle Structures and Unified Field Theory Based on Yangton and Yington Hypothetical Theory

Edward T. H. Wu

Solar Buster Corporation, Los Angeles, USA

Email address:
edwardthwu@yahoo.com

To cite this article:
Edward T. H. Wu. Subatomic Particle Structures and Unified Field Theory Based on Yangton and Yington Hypothetical Theory. *American Journal of Modern Physics*. Vol. 4, No. 4, 2015, pp. 189-195. doi: 10.11648/j.ajmp.20150404.15

Abstract: A hypothetical theory of a Yangton and Yington circulating pair with an inter-attractive Force of Creation is proposed as the "Origin of Creation". When this circulating pair travels in space it is known as a "Photon". Otherwise, it is known as a "Wu's Pair" also as "Still Photon", which makes the basic building block of all Matter. The structures of Quarks, Neutrinos, Graviton particles, Electrons, Positrons, Protons, Neutrons, and Dark Matter are proposed based on "Wu's Pairs". Also, it is assumed that Force of Creation is the only fundamental force in the Universe that could create and unify the Four Basic Forces. Gravitation is formed by the attractive force between two Graviton particles with String Structures that are made of "Wu's Pairs" in the same circulating direction. Electric Forces, either attractive or repulsive, can be formed between static Electrons, Protons and Positrons. Electromagnetic Force is generated between moving Electrons, Positrons and Protons. Two atoms both with single outer layer Electron are attractive to each other while circulating and spinning in the same direction and repulsive in the opposite direction. Both the Weak Force between a Neutron and a Positron as well as the Strong Force between two Neutrons or between a Proton and a Neutron is proposed and interpreted based on "Wu's Pairs" and Force of Creation.

Keywords: Yangton, Yington, Wu's Pairs, Photon, Electron, Proton, Neutron, Unified Field Theory.

Annex 3

American Journal of Modern Physics
2015; 4(6): 367-373
Published online November 10, 2015 (http://www.sciencepublishinggroup.com/j/ajmp)
doi: 10.11648/j.ajmp.20150406.12
ISSN: 2326-8867 (Print); ISSN: 2326-8891 (Online)

Light Speed in Vacuum Is not a Constant and Time Doesn't Change with Velocity – Discrepancies Between Relativities and Yangton & Yington Theory

Edward T. H. Wu

Solar Buster Corporation, Los Angeles, USA

Email address:
edwardthwu@yahoo.com

To cite this article:
Edward T. H. Wu. Light Speed in Vacuum Is not a Constant and Time Doesn't Change with Velocity – Discrepancies Between Relativities and Yangton & Yington Theory. *American Journal of Modern Physics*. Vol. 4, No. 6, 2015, pp. 367-373. doi: 10.11648/j.ajmp.20150406.12

Abstract: Light Speed in a vacuum, instead of being a constant, changes with those observers moving at different speeds and directions with respect to light origins. Time, on the other hand, instead of moving slower with the traveler, it always keeps the same rate. These facts disagree with Einstein's Special Relativity. Light speed in a vacuum is a constant only if it is observed from light origins and those positions stationary to light origins in Absolute Space System. This is because the emission of a Yangton and Yington circulating pair (Wu's Pair or Still Photon) from the surface of a matter (String Structure or Graviton) to form a free Photon traveling in vacuum is a Non-Inertia Transformation and it only requires a small fixed amount of Force of Separation. Since light speed is not a constant to those observers moving at different speeds and directions with respect to light origins, Velocity Time Dilation derived from Einstein's Special Relativity, is not true and could never exist. Absolute Space System, Vision of Light and Non-Inertia Transformation are introduced to explain the relationships between Space, Time and Relativities. The Doppler Effect, Blue Shift and Redshift are due to the Non-Inertia Transformation of light emission. Length contraction is caused by the difference of Visions of Light instead of Velocity Time Dilation. In an Inertia System, because of the same Visions of Light, the same light speeds in a vacuum can be observed by all observers. Furthermore, the Michelson - Morley Experiment proves that for two split light beams traveling in a vacuum, the same light speeds can also be observed. Time is the measurement of the cycles of a fundamental process from start to the end of an event. Both time and light speed at large Gravitational Field have relatively slower rates, which may be caused by the longer period and lower frequency of Yangton and Yington circulation due to the influence of a large Gravitational Field. This agrees with Gravitational Time Dilation in Einstein's General Relativity.

Keywords: Special Relativity, Light Speed, Vision of Light, Velocity Time Dilation, Gravitational Time Dilation, Length Contraction, Michelson-Morley, Yangton, Yington, Photon, Wu's Pair.

Annex 4

American Journal of Modern Physics
2016; 5(2): 20-24
http://www.sciencepublishinggroup.com/j/ajmp
doi: 10.11648/j.ajmp.20160502.11
ISSN: 2326-8867 (Print); ISSN: 2326-8891 (Online)

Gravitational Waves, Newton's Law and Coulomb's Law Interpreted by Particle Radiation and Interaction Theory Based on Yangton & Yington Theory

Edward T. H. Wu

Solar Buster Corporation, Los Angeles, United States

Email address:
edwardthwu@yahoo.com

To cite this article:
Edward T. H. Wu. Gravitational Waves, Newton's Law of Universal Gravitation and Coulomb's Law of Electrical Forces Interpreted by Particle Radiation and Interaction Theory Based on Yangton & Yington Theory. *American Journal of Modern Physics.* Vol. 5, No. 2, 2016, pp. 20-24. doi: 10.11648/j.ajmp.20160502.11

Received: March 4, 2016; **Accepted**: March 10, 2016; **Published**: March 30, 2016

Abstract: The structures of Photons, Gravitons, Electrons, and Positrons are proposed based on Wu's Pair, a Yangton and Yington circulating pair. The gravitational force is generated by the attractive force through close contact between two Gravitons of String Structures made of "Wu's Pairs" having the same circulation direction. Electrical Forces are formed by either attractive or repulsive forces via close contact between Electrons and Positrons/Protons. Graviton Radiation and Electron Radiation with Inverse Square Law and Contact Interactions, defined as "Particle Radiation and Interaction Theory", are proposed as the mechanisms to explain Newton's Law of Universal Gravitation and Coulomb's Law of Electrical Forces. Furthermore, Gravitational Waves instead of ripples of space-time, are in fact the fluctuation of the total gravitational forces of two merging Black Holes detected in Line of Sight by Earth observers. This can be interpreted by Graviton Radiation and Interaction Theory based on Yangton & Yington Theory.

Keywords: Yangton, Yington, Photon, Wu's Pair, Graviton, Electron, Positron, Gravitational Force, Electrical Forces, Graviton Radiation, Electron Radiation, Particle Radiation and Interaction, Newton's Law, Coulomb's Law, Gravitational Waves.

Annex 5

American Journal of Modern Physics
2016; 5(4): 45-50
http://www.sciencepublishinggroup.com/j/ajmp
doi: 10.11648/j.ajmp.20160504.11
ISSN: 2326-8867 (Print); ISSN: 2326-8891 (Online)

Mass, Momentum, Force and Energy of Photon and Subatomic Particles, and Mechanism of Constant Light Speed Based on Yangton & Yington Theory

Edward T. H. Wu

Solar Buster Corporation, Los Angeles, USA

Email address:
edwardthwu@yahoo.com

To cite this article:
Edward T. H. Wu. Mass, Momentum, Force and Energy of Photon and Subatomic Particles, and Mechanism of Constant Light Speed Based on Yangton & Yington Theory. *American Journal of Modern Physics*. Vol. 5, No. 4, 2016, pp. 45-50. doi: 10.11648/j.ajmp.20160504.11

Received: April 18, 2016; **Accepted**: April 27, 2016; **Published**: June 7, 2016

Abstract: The meaning of Mass, Momentum, Force and Energy of a substance is clearly defined. Their relationships to each other are discussed. The reason that a Photon has energy $E = h\nu$ and momentum $P = h/\lambda$ is explained. Light traveling in a vacuum at a constant speed $C = 3 \times 10^8$ m/sec while converting from a Wu's Pair through the separation and ejection process from its parent substance is discussed in detail based on Yangton and Yington Theory. In Black Body radiation, a Photon emitted at a high temperature, having smaller size, higher frequency and higher energy, is also explained.

Keywords: Mass, Force, Momentum, Energy, Yangton, Yington, Photon, Wu's Pair, Graviton, Electron, String Structure, Light Speed, Black Body Radiation.

Annex 6

American Journal of Modern Physics
2016; 5(4): 58-64
http://www.sciencepublishinggroup.com/j/ajmp
doi: 10.11648/j.ajmp.20160504.13
ISSN: 2326-8867 (Print); ISSN: 2326-8891 (Online)

Time, Space, Gravity and Spacetime Based on Yangton & Yington Theory, and Spacetime Shrinkage Versus Universe Expansion

Edward T. H. Wu

Solar Buster Corporation, Los Angeles, USA

Email address:
edwardthwu@yahoo.com

To cite this article:
Edward T. H. Wu. Time, Space, Gravity and Spacetime Based on Yangton & Yington Theory, and Spacetime Shrinkage Versus Universe Expansion. *American Journal of Modern Physics*. Vol. 5, No. 4, 2016, pp. 58-64. doi: 10.11648/j.ajmp.20160504.13

Received: May 12, 2016; **Accepted**: May 30, 2016; **Published**: July 13, 2016

Abstract: Wu's Pair is proposed as the building block of all matter, therefore Time and Length can be measured by the Period and Size of the circulation of a Wu's Pair. For a local event, object and process, the Time, Length and Velocity measured by the Period and Size of a Wu's Pair has equal values as that of the corresponding identical events, objects and processes occurring at different locations, measured by the corresponding Periods and Sizes of Wu's Pairs. However, for a remote event, object and process, the Time, Length and Velocity change with the Period and Size of a Wu's Pair at observation. For an event, object and process occurring in a large Gravitational Field and an ancient Universe, Time runs slower, Length gets longer and Velocity is observed smaller on Earth. As a result, light travels at a lower speed with lower frequency and larger wavelength from a large Gravitational Field known as "Gravitational Redshift"; also from a star of a few millions light years away known as "Cosmological Redshift".

Spacetime is a four dimensional system based on the Period and Size of a Wu's Pair. Since the Period and Size of Wu's Pairs change with Gravitational Field and Concentration of Gravitons,

Spacetime is a function of Gravitational Field and Concentration of Gravitons. In Wu's Pair, Time is proportional to 3/2 power of Length which is named as "Wu's Spacetime Theory". Furthermore, instead of the "Universe Accelerating Expansion Theory" which is based on the non-existing Dark Energy, "Spacetime Shrinkage Theory" is proposed to explain the accelerating expansion of the Universe.

Keywords: Yangton, Yington, Photon, Wu's Pair, Graviton, Particle Radiation, Gravitational Force, Gravitational Field, Time, Space, Spacetime, Black Hole, Gravitational Redshift, Cosmological Redshift, Universe Expansion, Dark Energy.

Annex 7

American Journal of Modern Physics
2017; 6(1): 10-15
http://www.sciencepublishinggroup.com/j/ajmp
doi: 10.11648/j.ajmp.20170601.12
ISSN: 2326-8867 (Print); ISSN: 2326-8891 (Online)

Redshift Caused by Acceleration Doppler Effect and Hubble's Law Based on Wu's Spacetime Accelerating Shrinkage Theory

Edward T. H. Wu

Solar Buster Corporation, Los Angeles, USA

Email address:
edwardthwu@yahoo.com

To cite this article:
Edward T. H. Wu. Redshift Caused by Acceleration Doppler Effect and Hubble's Law Based on Wu's Spacetime Accelerating Shrinkage Theory. *American Journal of Modern Physics*. Vol. 6, No. 1, 2017, pp. 10-15. doi: 10.11648/j.ajmp.20170601.12

Received: January 9, 2017; **Accepted**: January 18, 2017; **Published**: March 4, 2017

Abstract: The Acceleration Doppler Effect is introduced to explain the Redshift phenomenon by the Photon Inertia Transformation process. Hubble's Law and Cosmological Redshift are interpreted by Wu's Spacetime Shrinkage Theory. In addition, Wu's Laws of Spacetime of Wu's Pairs and Photons with respect to the age of universe and the gravitational field are derived and summarized.

Keywords: Yangton, Yington, Wu's Pair, Photon, Vision of Light, Doppler Effect, Acceleration Doppler Effect, Redshift, Cosmological Redshift, Hubble's Law, Universe Expansion, Spacetime Shrinkage.

Annex 8

IOSR Journal of Applied Physics (IOSR-JAP)
e-ISSN: 2278-4861.Volume 9, Issue 5 Ver. III (Sep. - Oct. 2017), PP 49-54
www.iosrjournals.org

Vision of Object, Vision of Light, Photon Inertia Transformation and Their Effects on Light Speed and Special Relativity

Edward T. H. Wu
Corresponding Author: Edward T. H. Wu

***Abstract**: The vision of an object, in spite of observed directly at the observation point, can be constructed by integrating the images of the object observed at a reference point during a period of time. Similarly, the vision of a photon (Vision of Light) observed at an observation point can also be constructed from the images of the photon observed at the light origin. When a photon emitted from a light source, it travels under two influences, ejection motion and inertia motion. On one hand, photon travels at a constant Absolute Light Speed (3×10^8 m/s) in its trajectory because of the constant ejection force caused in the photon emission process; one the other hand, it is dragged into a direction and speed the same as that of the light source due to Photon Inertia Transformation. In other words, light speed is a vector summation of Absolute Light Speed and the speed of light source observed at the observation point. Light speed in an inertia system is always a constant (unnecessarily 3×10^8 m/s) because the same Vision of Light can be observed at all stationary positions in the inertia system. However, oppose to Einstein's Special Relativity Theory that the light speed is always a constant no matter the light sources and observers, light speed in fact changes with the observers at different moving speeds and directions. As a result, Einstein's Special Relativity and Velocity Time Dilation theories are false and time doesn't change with velocity at all.*

***Keywords**: Vision of Object, Vision of Light, Photon Inertia Transformation, Light Speed, Special Relativity, Velocity Time Dilation*

Date of Submission: 20-09-2017 Date of acceptance: 07-10-2017

Annex 9

IOSR Journal of Applied Physics (IOSR-JAP)
e-ISSN: 2278-4861.Volume 10, Issue 1 Ver. I (Jan. – Feb. 2018), PP 58-62
www.iosrjournals.org

Hubble's Law Interpreted by Acceleration Doppler Effect and Wu's Spacetime Reverse Expansion Theory

Edward T. H. Wu
Corresponding Author: Edward T. H. Wu

Abstract: *Hubble's Law is an experimental result that can be applied to a star with linear relation between recessional velocity and Redshift, subject to that the star is at more than 5 billion light years away and is moving away from earth at an acceleration speed faster than the light speed. Although Acceleration Doppler Effect can be used to derive Hubble's Law, an imaginary Dark Energy is suggested to explain the acceleration together with an expansion theory of the universe to support the super fast moving speed. To avoid these problems, a Spacetime Shrinkage model is proposed and a Spacetime Reverse Expansion Theory is successfully derived to interpret Hubble's Law.*

Keywords*: Hubble's Law, Dark Energy, Doppler Effect, Acceleration Doppler Effect, Universe Expansion, Wu's Spacetime, Yangton and Yington, Wu's Pairs, Spacetime Shrinkage, Reverse Expansion*

Date of Submission: 12-01-2018 Date of acceptance: 03-02-2018

Annex 10

IOSR Journal of Applied Physics (IOSR-JAP)
e-ISSN: 2278-4861.Volume 10, Issue 2 Ver. I (Mar. – Apr. 2018), PP 13-21
www.iosrjournals.org

Wu's Spacetime Field Equation Based On Yangton And Yington Theory

Edward T. H. Wu
Corresponding Author: Edward T. H. Wu

Abstract: *Wu's Spacetime Field Equation is derived from Yangton and Yington Theory based on Wu's Unit Length l_{yy} (the diameter of Yangton and Yington Circulating Pairs) and Wu's Unit Time t_{yy} (the period of Yangton and Yington Circulating Pairs). Wu's Unit Length and Wu's Unit Time are correlated to each other by Wu's Spacetime Theory. They are also dependent on the gravitational field and the aging of the universe. Furthermore, instead of being a constant, the speed of light C is a function of Wu's Unit Length l_{yy}, which can increase the acceleration (the curvature of Spacetime) to form a deep continuum in Spacetime along the edge of a spherical mass (or black hole). As a result, the existence of black hole can be interpreted by Wu's Spacetime Field Equation. Also, the expansion of the universe can be explained by Wu's Spacetime Shrinkage Theory and Wu's Spacetime Reverse Expansion Theory without the modification of Wu's Spacetime Field Equation with Einstein's Cosmological Constant and dark energy.*
Keywords: *General Relativity. Einstein's Field Equations, Yangton and Yington, Wu's Pairs, Spacetime, Spacetime Shrinkage, Universe Expansion, Redshift, Black Hole, Dark Energy, Cosmological Constant.*

Date of Submission: 22-02-2018 Date of acceptance: 10-03-2018

Annex 11

IOSR Journal of Applied Physics (IOSR-JAP)
e-ISSN: 2278-4861.Volume 10, Issue 4 Ver. I (Jul. – Aug. 2018), 50-56
www.iosrjournals.org

Standard Model and Quantum Field Theory versus Wu's Pairs and Yangon and Yangon Theory

Edward T. H. Wu
Corresponding Author: Edward T. H. Wu

[Abstract]: *Standard Model is a group of subatomic particles derived from a mathematical model based on quantum field theory and Yang Mills Theory. In contrast, Wu's Pairs, a physical model are proposed as the building blocks of all subatomic particles based on the Yangton and Yington Theory. In this paper, some critical issues of quantum field theory are discussed, such as "What are the subatomic particles made of?", "What are the symmetries of gluons?", "What are the Higgs Bosons and Higgs field?" and "What the quantum fields really are?" As a result, three innovative theories "Quantum Gravity Theory based on the gravitons of a string structure and the theory of particle radiation and contact interaction", "Quantum Fields based on the distributions of contact interactions caused by particle radiations" and "Unified field theory based on the string structures made of Wu's Pairs and Force of Creation" are proposed.*

[Keywords]: *Standard Model, Quantum Field Theory, Wu's Pairs, Yangton and Yington, String Theory, Unified Field Theory, Graviton, Quantum Gravity, Higgs Boson, Particle Radiation*

Date of Submission: 29-06-2018 Date of acceptance: 16-07-2018

Annex 12

IOSR Journal of Applied Physics (IOSR-JAP)
e-ISSN: 2278-4861.Volume 10, Issue 4 Ver. II (Jul. – Aug. 2018), 17-21
www.iosrjournals.org

Five Principles of the Universe and the Correlations of Wu's Pairs and Force of Creation to String Theory and Unified Field Theory

Edward T. H. Wu

Abstract: *Yangton and Yington Theory is based on Yangton and Yington circulating particle pairs (Wu's Pairs) with a build-in inter-attractive force (Force of Creation) that is proposed as the fundamental building blocks of the universe. Although Yangton and Yington Theory is only a hypothetical theory, the whole concept was developed based on the Five Principles of the Universe. In this paper, these principles are discussed in detail. Also, String Theory is explained by the string structures built upon Wu's Pairs and Unified Field Theory is interpreted by the subatomic structures and their corresponding four basic forces induced from Force of Creation.*

Keywords: Wu's Pairs, Yangton and Yington, Force of Creation, Subatomic Particles, String Theory, Four Basic Forces, Unified Field Theory, General Relativity, Quantum Field Theory, Quantum Gravity

---

Date of Submission: 14-07-2018 Date of acceptance: 31-07-2018

Annex 13

IOSR Journal of Applied Physics (IOSR-JAP)
e-ISSN: 2278-4861.Volume 10, Issue 5 Ver. I (Sep. – Oct. 2018), 80-84
www.iosrjournals.org

Mass, Time, Length, Vision of Object and Principle of Correspondence Based on Yangton and Yington Theory

Edward T. H. Wu
Corresponding Author: Edward T. H. Wu

Abstract: *The meanings of mass, time and length are discussed. They all have their own absolute values which don't change with the measurements. But the "unit" and "amount" of the units can be different subject to each measurement method. The same "unit" of the same measurement can also change its value subject to the gravitational force and aging of the universe. In addition, because the vision of object can change with the relative speeds and directions between the object and the observer, the distance of a moving object can also be different subject to each observation. However, the relative length and relative time are always constants when the corresponding identical objects and corresponding identical events are measured by the corresponding identical unit length and corresponding identical unit time. Principle of Correspondence is proposed which implies that all physical laws should maintain the same in the inertial systems measured by the corresponding identical units.*

Keywords: *Mass, Time, Length, Wu's Pairs, Relativity, Velocity Time Dilation, Yangton and Yington, Vision of Object. Theory of Vision, Corresponding Identical Object, Corresponding Identical Event, Principle of Correspondence.*

Date of Submission: 03-09-2018 Date of acceptance: 18-09-2018

Annex 14

IOSR Journal of Applied Physics (IOSR-JAP)
e-ISSN: 2278-4861.Volume 10, Issue 5 Ver. II (Sep. - Oct. 2018), 45-50
www.iosrjournals.org

A Summary of Yangton and Yington Theory and Their Interpretations on Subatomic Particles, Gravitation and Cosmology

Edward T. H. Wu
Corresponding Author: Edward T. H. Wu

Abstract: *Yangton and Yington Theory is a hypothetical theory based on Yangton and Yington circulating particle pairs (Wu's Pairs) with a build-in inter-attractive force (Force of Creation) that is proposed as the fundamental building blocks of the universe. It has explained successfully the formation of all subatomic particles, propagation of gravitational force, cosmological redshift, universe expansion and many other correlations between space, time, energy and matter. More than 31 major physical phenomena and theories are interpreted and derived from Yangton and Yington Theory. In summary, a road map of systematic derivation and a correlation network between the major physical phenomena and Yangton and Yington Theory are presented.*

Keywords: *Wu's Pairs, Yangton andYington, Force of Creation, Subatomic Particles, String Force, String Theory, Quantum Field Theory, Quantum Gravity Theory, Unified Field Theory, Standard Model, Dark Matter, Antimatters, Graviton, Particle Radiation, Gravitational Waves, Photon, Special Relativity, General Relativity, Light Speed, Doppler Effect, Acceleration Doppler Effect, Inertia Transformation, Redshift, Cosmological Redshift, Gravitational Redshift, Vision of Light, Vision of Object, Principle of Correspondence, Spacetime, Field Equations, Dark Energy, Universe Expansion, Reverse Expansion, Spacetime Shrinkage Theory.*

Date of Submission: 23-09-2018 Date of acceptance: 08-10-2018

Annex 15

IOSR Journal of Applied Physics (IOSR-JAP)
e-ISSN: 2278-4861.Volume 11, Issue 1 Ser. II (Jan. – Feb. 2019), PP 03-07
www.iosrjournals.org

Hubble's Law Derived from Wu's Spacetime Shrinkage Theory and Wu's Spacetime Reverse Expansion Theory versus Universe Expansion Theory

Edward T. H. Wu

Abstract: *Stars that are more than 5 billion light years away from earth obey Hubble's Law having linear relations between redshift, recessional velocity and proper distance. Although Acceleration Doppler Effect can be used to derive Hubble's Law based on a superfast acceleration speed and an imaginary Dark Energy for the interpretation of the universe expansion theory. Wu's Spacetime Shrinkage Theory based on the shrinkage of the diameter l_{yy} (Wu's Unit Length) and circulation period t_{yy} (Wu's Unit Time) of Wu's Pairs – the building blocks of the universe due to the aging of the universe is successfully used to derive Hubble's Law without acceleration and Dark Energy. Because of these reasons, Wu's Spacetime Reverse Expansion Theory is proposed for a better explanation of Cosmological Redshift and Hubble's Law.*
Keywords: *Hubble's Law, Dark Energy, Doppler Effect, Acceleration Doppler Effect, Cosmological Redshift, Universe Expansion, Wu's Spacetime, Yangton and Yington, Wu's Pairs, Spacetime Shrinkage, Reverse Expansion*

Date of Submission: 25-01-2019 Date of acceptance: 07-02-2019

Annex 16

IOSR Journal of Applied Physics (IOSR-JAP)
e-ISSN: 2278-4861.Volume 11, Issue 2 Ser. I (Mar. – Apr. 2019), PP 57-61
www.iosrjournals.org

Einstein's $E = MC^2$ as Energy Conversion Instead of Mass and Energy Conservation and Energy and Space Annihilation Based on Yangton and Yington Theory

Edward T. H. Wu

Abstract: *Einstein's $E = MC^2$ is not a law of mass and energy conservation. Instead, it is only an energy conversion between matter's structure energy and photon's kinetic energy. On the other hand, Wu's Pairs are created by the energy generated in big bang explosion, which is a typical mass and energy conversion. According to Yangton and Yington Theory, in the beginning, energy and space are first generated from nothing, and then matter and time are induced from energy. At the end of the universe, to reverse the process, matter will convert to energy first, and then energy and space annihilation will happen either in black hole or through aging of the universe.*

Keywords: *Yangton and Yington, Wu's Pairs, Force of Creation, Subatomic Particles, Energy and Mass Conservation, Black Hole, Big Bang, Singularity, Spacetime Shrinkage, Antiparticle Annihilation, Energy and Space Annihilation.*

Date of Submission: 26-02-2019 Date of acceptance: 12-03-2019

Annex 17

IOSR Journal of Applied Physics (IOSR-JAP)
e-ISSN: 2278-4861.Volume 11, Issue 2 Ser. III (Mar. – Apr. 2019), PP 13-18
www.iosrjournals.org

Einstein's Spacetime and Einstein's Field Equations Versus Wu's Spacetime and Wu's Spacetime Field Equations

Edward T. H. Wu

Abstract: *Wu's Spacetime is a four dimensional system based on Wu's Unit Length l_{yy} and Wu's Unit Time t_{yy} which are related to each other by Wu's Spacetime Theory $t_{yy} = \gamma l_{yy}^{3/2}$. Einstein's Spacetime is a special Wu's Spacetime based on earth. According to Yangton and Yington Theory, Wu's Unit Length l_{yy} on a massive star is much bigger than l_{yy0} on earth. Because $a \infty C^4 \infty l_{yy}^{-2}$, the Amount of Normal Unit Acceleration "a" measured on the star is much bigger than "a_0"measured on earth. In other words, for a massive star, Wu's Spacetime Field Equation measured on the star has much deeper slope (curvature) than that of Einstein's Field Equation measured on earth. Furthermore, because of the large Wu's Unit Length l_{yy} caused by the huge gravitational force, a hollow structure in the center of a black hole is expected. Also because of the Photon Inertia Transformation and the large acceleration in the center of a black hole based on Wu's Spacetime Field Equations, it is predicted that photon can be trapped inside the event horizon of a black hole.*

Keywords: *Theory of Correspondence, Field Equation, Einstein's Field Equation, Cosmological Constant, Spacetime, Yangton and Yington, Wu's Pairs, Wu's Spacetime Theory, Wu's Spacetime Field Equation, Black Hole.*

Date of Submission: 05-04-2019 Date of acceptance: 20-04-2019

Annex 18

IOSR Journal of Applied Physics (IOSR-JAP)
e-ISSN: 2278-4861.Volume 11, Issue 3 Ser. I (May. – June. 2019), PP 15-17
www.iosrjournals.org

Einstein's Seven Mistakes

Edward T. H. Wu
Corresponding Author: Edward T. H. Wu

Abstract: *Einstein derived his theories including Special Relativity, General Relativity, Spacetime, Field Equations and Mass and Energy Conservation, based on two wrong assumptions: (a) Light speed is always constant no matter the light source and observer, and (b) Acceleration is the principle factor of Spacetime. In contrast, according to Yangton and Yington Theory, it is realized that (a) Light speed is not constant, instead, it is the vector summation of Absolute Light Speed C ($C \propto l_{yy}^{-1/2}$) and Inertia Light Speed, and (b) Acceleration is not a principle factor, instead, gravitational field and aging of the universe are the principle factors of Wu's Spacetime. In fact, the time and length of an object and event are a function of the Wu's Unit Time (t_{yy}) and Wu's Unit Length (l_{yy}) depending on the gravitational field and the aging of the universe no matter of the acceleration. Einstein's Spacetime is the solution of Einstein's Field Equations. It is a twisted system observed on earth with coordinates of the object and event based on a twisted unit time, unit length and Absolute Light Speed (3×10^8 m/s). In contrast, Wu's Spacetime on earth is a straight system with coordinates of the object and event depending on their corresponding Wu's Unit Time t_{yy} and Wu's Unit Length l_{yy} measured by Wu's Unit Time t_{yy0}, Wu's Unit Length l_{yy0} and Absolute Light Speed C_0 (3×10^8 m/s) on earth.*

Keywords: *Light Speed, Special Relativity, Velocity Time Dilation, Relativistic Mass, Relativistic Length, General Relativity, Field Equation, Einstein's Field Equation, Spacetime, Mass and Energy Conservation, Yangton and Yington, Wu's Pairs, Wu's Spacetime, Wu's Spacetime Theory, Wu's Spacetime Field Equations.*

Date of Submission: 18-04-2019 Date of acceptance: 04-05-2019

Annex 19

IOSR Journal of Applied Physics (IOSR-JAP)
e-ISSN: 2278-4861.Volume 11, Issue 3 Ser. III (May. – June. 2019), PP 41-45
www.iosrjournals.org

General Relativity and Yangton and Yington Theory – Corresponding Identical Object and Event in Large Gravitational Field Observed on Earth

Edward T. H. Wu

Abstract: *According to Yangton and Yington Theory, for an observation on earth, in addition to have larger length and time, an object and event happen on a massive star (black hole) with large gravitational field has smaller velocity and acceleration comparing to that of the corresponding identical object and event happen on earth with less gravitational field. This result agrees very well with general relativity.*
Keywords: *General Relativity, Time Dilation, Yangton and Yington, Wu's Pairs, Wu's Unit Time, Wu's Unit Length, Principle of Correspondence, Corresponding Identical Object, Corresponding Identical Event, Wu's Spacetime Theory.*

Date of Submission: 02-06-2019 Date of acceptance: 17-06-2019

Annex 20

IOSR Journal of Applied Physics (IOSR-JAP)
e-ISSN: 2278-4861.Volume 11, Issue 4 Ser. II (Jul. – Aug. 2019), PP 01-06
www.iosrjournals.org

The correlations of Wu Constant and Wu's Spacetime Constant to Hubble Constant

Edward T. H. Wu

[Abstract]: *Wu Constant K and Wu's Spacetime Constant γ are derived from Yangton and Yington Circulation model. Also, Hubble's Law and Hubble Constant H_0 are derived from Spacetime Shrinkage Theory based on the aging of the Universe. This is called Spacetime Reverse Expansion Theory which successfully explains the expansion of the universe without the conflict of dark energy. Furthermore, Wu Constant K and Wu's Spacetime Constant γ can be expressed in terms of Hubble Constant H_0.*

[Keywords] *Hubble's Law, Yangton and Yington, Wu's Pairs, Wu Constant, Wu's Spacetime, Wu's Spacetime Theory, Wu's Spacetime Constant, Spacetime Shrinkage, Acceleration Doppler Effect, Dark Energy, Universe Expansion.*

Date of Submission: 13-07-2019 Date of acceptance: 29-07-2019

Annex 21

IOSR Journal of Applied Physics (IOSR-JAP)
e-ISSN: 2278-4861.Volume 11, Issue 5 Ser. I (Sep. – Oct. 2019), PP 68-72
www.iosrjournals.org

Derivations of Planck Constant, De Broglie Matter Waves and Mass of Photon (Wu's Pair) from Yangton and Yington Theory

Edward T. H. Wu

Abstract: *Spinning particles are simulated by a whirlpool model. Because the momentum P of the spinning particle is proportional to the mass m and the spin frequency v of the particle, therefore* $P = Kmv = KmC/\lambda = mh_0/\lambda$ *(where* h_0 *is named General Planck Constant), De Broglie Wavelength* $\lambda = mh_0/P$*, and Planck constant* $h = m_{yy}h_0$ *(where* m_{yy} *is the mass of a photon or a Wu's Pair). In addition, De Broglie wavelength, momentum and energy of the electron in Bohr Model are derived. As a result, all Planck constant h in the old formula are replaced by* $(m_e/m_{yy})h$ *such that in the new version, De Broglie wavelength* $\lambda = (m_e/m_{yy})h/P$*, momentum* $P = m_e(KZe^2)/(n\hbar)(m_e/m_{yy})$ *and energy* $E = -\frac{1}{2}m_e(KZe^2)^2/(n^2\hbar^2)(m_e/m_{yy})^2$*. Furthermore, the mass of photon (Wu's Pair) can be derived as* $m_{yy} = h/h_0 = h/KC$ *(where* h_0 *is General Planck Constant and K is whirlpool constant).*

Keywords: *De Broglie Waves, Matter Waves, Wave Particle Duality, Photon, Electron, Spin, Planck Constatnt, Yangton and Yington, Wu's Pairs.*

Date of Submission: 18-09-2019 Date of Acceptance: 03-10-2019

Annex 22

IOSR Journal of Applied Physics (IOSR-JAP)
e-ISSN: 2278-4861.Volume 11, Issue 6 Ser. II (Nov. – Dec. 2019), PP 51-55
www.iosrjournals.org

Higgs Boson and Graviton Interpreted by String Force and String Structures Based on Wu's Pairs and Yangton and Yington Theory

Edward T. H. Wu

Abstract: Standard Model is a group of subatomic particles derived from a mathematical model based on quantum field theory and Yang Mills Theory. In contrast, Wu's Pairs, a physical model are proposed as the building blocks of all subatomic particles based on the Yangton and Yington Theory. Since Higgs Bosons can be considered as the carriers of string force that are generated by Wu's Pairs, therefore the magnitude of the barrier caused by the string force carried by Higgs Bosons is proportional to the amount of Wu's Pairs. In other words, the mass of a particle is proportional to the amount of Higgs Bosons as that of Wu's Pairs. This concurs with that the mass is the total amount of Wu's Pairs based on Yangton and Yington Theory. When two string structures come together, they can attract to each other either end to end or side by side. These attractive only forces are known as "Gravitational Force" and the string structures that produce the gravitational force are called "Gravitons".
Keywords: Standard Model, Quantum Field Theory, Wu's Pairs, Yangton and Yington, String Theory, Unified Field Theory, Graviton, Quantum Gravity, Higgs Boson, Particle Radiation.

Date of Submission: 22-11-2019 Date of Acceptance: 06-12-2019

Annex 23

IOSR Journal of Applied Physics (IOSR-JAP)
e-ISSN: 2278-4861.Volume 11, Issue 6 Ser. II (Nov. – Dec. 2019), PP 58-61
www.iosrjournals.org

Event Horizon and Black Hole Interpreted by Photon Inertia Transformation and Yangton and Yington Theory

Edward T. H. Wu

[Abstract]: *When a light source is accelerating towards the center of a black hole, because of the Photon Inertia Transformation, the emitted photon compiles two competing speeds in opposite directions: (1) outward Absolute Light Speed and (2) inward Inertia Light Speed. As a result, Event Horizon is the place where photon is in idle with zero net speed. Inside the Event Horizon, photon moves further down to the center of the black hole and it can never escape. Black Hole is generated by a massive gravitational force. Since Wu's Pair expands with gravitational force such that a hollow structure can be formed inside the Black Hole. At the center of the Black Hole, a singularity is formed where the circulation of Wu's Pairs are broken down such that Yangton and Yington could recombine and destroy each other to become None. Black Hole is the grave yard of all matter.*

[Keywords]: *Wu's Pairs, Yangton and Yington, Photon Inertia Transformation, Absolute Light Speed, Inertia Light Speed, Light Speed Equation, Black Hole, Event Horizon, Aging of the Universe.*

Date of Submission: 28-11-2019 Date of Acceptance: 13-12-2019

Annex 24

IOSR Journal Of Applied Physics (IOSR-JAP)
e-ISSN: 2278-4861.Volume 12, Issue 1 Ser. III (Jan. – Feb. 2020), PP 09-19
www.Iosrjournals.Org

A Summary of Wu's Spacetime Field Equation and Its Comparison to Einstein's Field Equation

Edward T. H. Wu

[Abstract]: *A summary of Wu's Spacetime Field Equation and some related subjects are reviewed, including Wu's Pairs, Wu's Spacetime, Wu's Spacetime Theory, Wu's Spacetime Shrinkage Theory, Time & Length, Principle of Correspondence, Distribution of Wu's Unit Length, Amount of Wu's Unit Quantities Measured At Reference Point, Field Equation, Wu's Spacetime Field Equation, Einstein's Spacetime and Einstein's Field Equation. A detailed comparison between Wu's Spacetime Equation and Einstein's Field Equation are discussed. Because the same term* GC_0^{-4} *appears in both equations, Einstein's Field Equation and Wu's Spacetime Field Equation observed on earth look like equivalent. However, there is no gravitational force in Einstein's Spacetime Field Equation. Acceleration is derived from the curvature of space-time continuum, which reflects the virtual distribution of matter and energy in the universe. On the other hand, in Wu's Spacetime Field Equation, matter does exist, as is the gravitational force. And the acceleration is indeed caused by the gravitational force.*

[Keywords]: *Wu's Pairs, Yangton and Yington, Wu's Spacetime, Wu's Spacetime Theory, Wu's Spacetime Shrinkage Theory, Principle of Correspondence, Gravitational Field, Field Equation, Spacetime, Wu's Spacetime Field Equation, Einstein's Field Equation. Graviton Radiation, Cosmological Redshift, Gravitational Redshift.*

Date of Submission: 29-01-2020 Date of Acceptance: 14-02-2020

Annex 25

IOSR Journal Of Applied Physics (IOSR-JAP)
e-ISSN: 2278-4861.Volume 12, Issue 1 Ser. III (Jan. – Feb. 2020), PP 20-26
www.Iosrjournals.Org

Perihelion Precession of Mercury and Deflection of Light Interpreted by Yangton and Yington Theory

Edward T. H. Wu

[Abstract]: *According to Yangton and Yington Theory, a moving corresponding identical object and event has velocity related to Wu's Unit Length by $V \propto l_{yy}^{-1/2}$. In addition, l_{yy} increases when gravitational force becomes bigger based on Wu's Spacetime Shrinkage Theory. Therefore, a moving corresponding identical object and event, its velocity decreases with the increase of gravitational force. Furthermore, like the light refraction on the surface of a transparent substance, the decreasing speed of the object and event can make its traveling path bent towards the massive mass. As a result, instead of the cured Spacetime proposed by Einstein's general relativity, deflection of light and Perihelion Precession of Mercury can be interpreted by Yangton and Yington Theory.*

[Keywords]: *Wu's Pairs, Yangton and Yington, Wu's Spacetime, Wu's Spacetime Theory, Wu's Spacetime Shrinkage Theory, Principle of Correspondence, General Relativity, Einstein's Spacetime, Cosmological Redshift, Gravitational Redshift, Expansion of Universe, Deflection of Light, Perihelion Precession of Mercury.*

Date of Submission: 29-01-2020 Date of Acceptance: 14-02-2020

Annex 26

Pure and Applied Mathematics Journal
2015; 4(3): 75-79
Published online April 22, 2015 (http://www.sciencepublishinggroup.com/j/pamj)
doi: 10.11648/j.pamj.20150403.13
ISSN: 2326-9790 (Print); ISSN: 2326-9812 (Online)

Refined Definitions in Real Numbers and Vectors and Proof of Field Theories

Edward T. H. Wu

DaVinci International Academy, Los Angeles, USA

Email address:
edwardthwu@yahoo.com

To cite this article:
Edward T. H. Wu. Refined Definitions in Real Numbers and Vectors and Proof of Field Theories. *Pure and Applied Mathematics Journal*. Vol. 4, No. 3, 2015, pp. 75-79. doi: 10.11648/j.pamj.20150403.13

Abstract: A set of new and refined principles and definitions in Real Numbers and Vectors are presented. What is a Vector? What is the meaning of the Addition of two Vectors? What is a Real Number? What is the meaning of their signs? What is the meaning of the Addition of two Real Numbers? What is the Summation Principle in Addition Operation? What is the Cancellation Principle in Addition Operation? What is the Meaning of the Multiplication of two Real Numbers? Is Field Theory a law? Can it be proven? All these issues are addressed in this paper. With better pictures and graphic presentations, proof of Field Theories in Real Numbers and Vectors including Commutativity, Associativity and Distributivity are also proposed.

Keywords: Vector, Real Number, Number Line, Number Vector, Number Space, Field Theory, Commutativity, Associativity, Distributivity.

Annex 27

Pure and Applied Mathematics Journal
2015; 4(4): 147-154
Published online June 17, 2015 (http://www.sciencepublishinggroup.com/j/pamj)
doi: 10.11648/j.pamj.20150404.12
ISSN: 2326-9790 (Print); ISSN: 2326-9812 (Online)

Mathematical Methodologies in Physics and Their Applications in Derivation of Velocity and Acceleration Theories

Edward T. H. Wu

Davinci International Academy, Los Angeles, California, USA

Email address:
edwardthwu@yahoo.com

To cite this article:
Edward T. H. Wu. Mathematical Methodologies in Physics and Their Applications in Derivation of Velocity and Acceleration Theories. *Pure and Applied Mathematics Journal*. Vol. 4, No. 4, 2015, pp. 147-154. doi: 10.11648/j.pamj.20150404.12

Abstract: The principles to use variables and mathematical methodologies in physics are addressed. A set of refined definitions, with designated variables, are used to derive the Velocity and Acceleration Theories in Distance Field and Vector Space. Mathematical methodologies such as Linear Algebra and Vector Calculus are used systematically in a step by step derivation process. The proof of the theories can be easily achieved by substitution of the designated variables with a set of parameters that matches the same assumptions and conditions in every step of the derivation process.

Keywords: Variables, Parameters, Velocity, Acceleration, Linear Algebra, Vector Calculus, Mathematical Methodology.

Annex 28

Nature Quantities and Measured Quantities

1. Nature Quantity

   Nature quantities are the nature forms of the physical quantities such as time and distance. They have no unit and amount except the physical values and relative comparisons. There are two typs of nature quantities, elementary nature quantity and composite nature quantity. For examples:

A. Elementary Nature Quantities

Distance	X
Time	T
Mass	M

B. Composite Nature Quantities

Velocity = ΔDistance/ΔTime	$V = \Delta X/\Delta T$
Acceleration = ΔVelocity/ΔTime	$A = \Delta V/\Delta T$
Force = Mass x Acceleration	$F = M\ A$

2. Measured Quantity

   Measured quantities are the physical quantities measured by human. It contains units and amounts. Like nature quantities, there are two typs of measured quantities, elementary measured quantity and composite measured quantity. For examples:

A. Elementary Measured Quantities

Distance	X meter
Time	T second
Mass	M kilogram

Where X, T and M are amounts in numbers. Meter, second and kilogram are elementary units depending on gravitational field and aging of the universe.

B. Composite Measured Quantities

Velocity = ΔDistance/ΔTime $V = \Delta X/\Delta T$ m/s

Acceleration = ΔVelocity/ΔTime $A = \Delta(\Delta V/\Delta T)/\Delta T$ m/s^2

Force = Mass x Acceleration $F = M A = M\Delta(\Delta V/\Delta T)/\Delta T$ kg m/s^2

Where $\Delta X/\Delta T$, $\Delta(\Delta V/\Delta T)/\Delta T$, $M\Delta(\Delta V/\Delta T)/\Delta T$ are amounts in numbers. m/s, m/s^2 and kg m/s^2 are composite units depending on gravitational field and aging of the universe.

Annex 29

Physical Meanings of Arithmetic Operations

1. Addition: The quantity of the summation of two quantities.

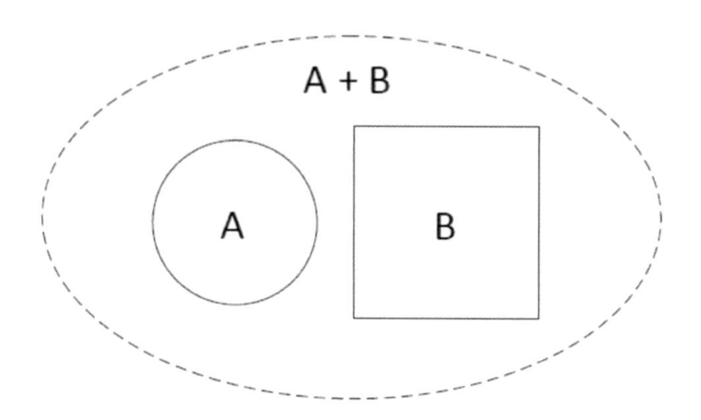

2. Subtraction: The quantity of subtraction of one quantity from the other quantity.

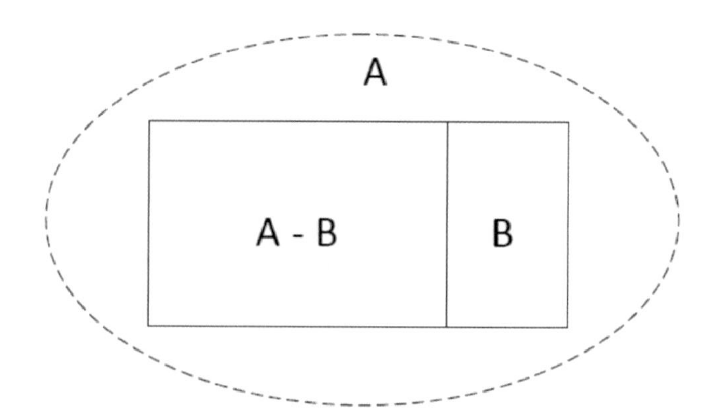

3. Multiplication: The quantity of the interaction between two quantities.

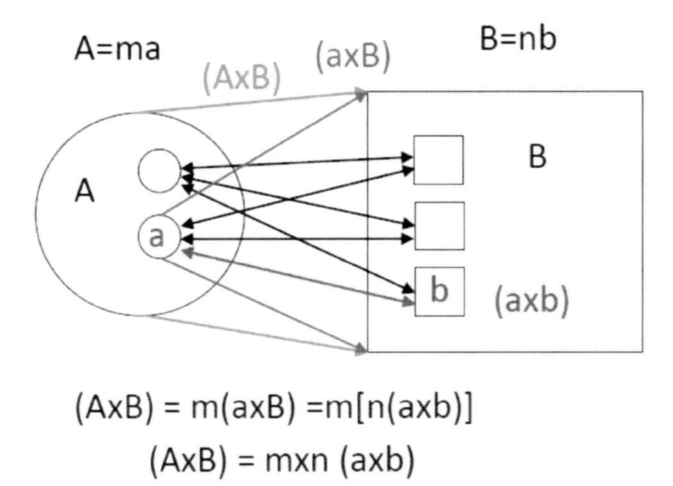

4. Division: The quantity of one quantity evenly distributed over the other quantity.

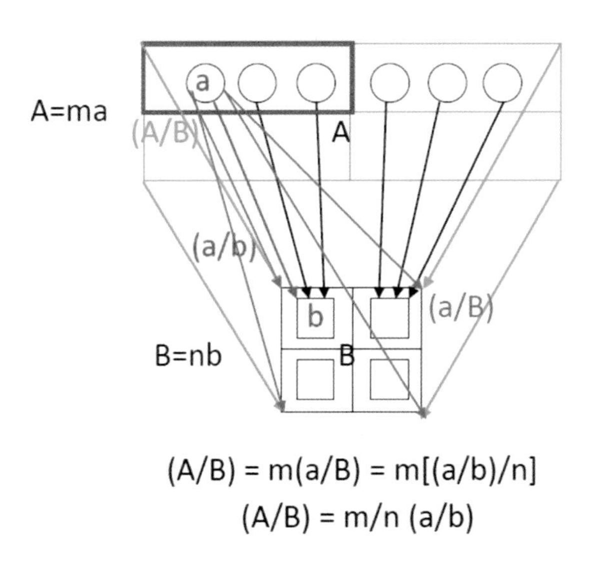

Annex 30

Constants and Constant Quantities in Physics

Quantity can be measured by the Amount of Unit multiplied by the Unit.

$$\text{Quantity} = \text{Amount} \times \text{Unit}$$

Two quantities X_1 and X_2 can be correlated to each other by a quantity called transformer T_{12}.

$$X_1 = T_{12} X_2$$

If two quantities are proportional to each other, then the transformer is a constant quantity K.

$$X_1 = K X_2$$

For example, the total force F is proportional to mass M and acceleration A, therefore, F can be represented as follows:

$$F \infty M \text{ (When A is a constant)}$$

$$F \infty A \text{ (When M is a constant)}$$

$$F = K (MA)$$

Also, a physical law or definition can commonly be represented by a formula (equation) having one quantity correlated to the other quantities with a constant quantity K or a constant number such as 1, π, etc.

For example, the gravitational force F can be derived as follows:

Because

$$F \infty M_1$$

$$F \infty M_2$$

$$F \infty 1/R^2$$

Therefore,

$$F = G\ M_1M_2/R^2$$

And

$$G = F\ R^2/M_1M_2$$

For Normal Units M_n (Kg), L_n (Meter) and F_n (Newton),

Because

$$fF_n = G\ m_1M_nm_2M_n/r^2L_n{}^2$$

Therefore,

$$G = fF_n\ r^2L_n{}^2/m_1M_nm_2M_n$$

Where f, r, m_1 and m_2 are amounts of units.

In fact, G can be measured on earth as follows:

$$G = f_0F_{n0}\ r_0{}^2L_{n0}{}^2/m_1M_nm_2M_n$$

$$G = (f_0\ r_0{}^2/m_1\ m_2)\ (F_{n0}\ L_{n0}{}^2/M_n{}^2)$$

Where G is the gravitational constant, f_0 is the Amount of Normal Unit Force (F_{n0}) on earth, r_0 is the Amount of Normal Unit Length (L_{n0}) on earth, m_1 and m_2 are the Amounts of Normal Unit Mass (M_n) of the two objects (Normal Unit Force and Normal Unit Length are functions of gravitational field and aging of the universe, but the Normal Unit Mass which has a fixed amount of Wu's Pairs). With a precision measurement, G has a value equals to 6.674×10^{11} N m^2 kg^{-2}.

Annex 31

Refraction and Deflection of Light

When light beam passes the interface of two transparent objects, because of the different light speeds, the path of the light beam bends to maintain the same frequency and the coherency. This phenomenon is called “Refraction of Light”.

Similarly, when a light beam travels close to a massive star, gravitational field becomes extremely large which makes Wu’s Unit Length l_{yy} bigger and Absolute Light Speed C smaller ($C \propto l_{yy}^{-1/2}$). As a result, the path of the light beam bends toward the star in order to maintain the same frequency and the coherency. This phenomenon is called “Deflection of Light”.

A schematic explanation can be presented as follows:

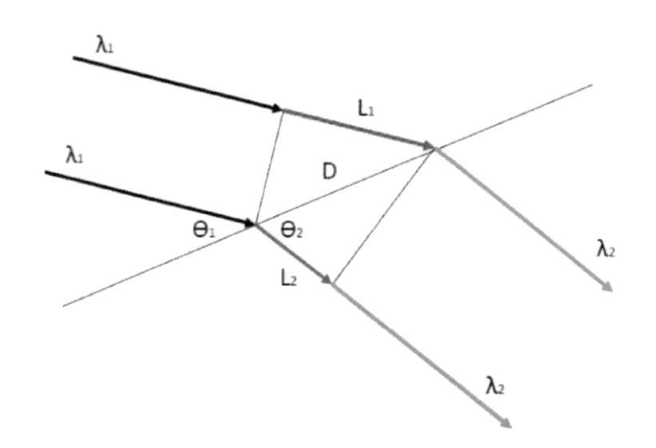

Because

$L_1 = V_1 t = D \cos \Theta_1$

$L_2 = V_2 t = D \cos \Theta_2$

$$V_1 > V_2$$

Therefore,

$L_1 > L_2$

$\cos \Theta_1 > \cos \Theta_2$

$$\Theta_1 < \Theta_2$$

Annex 32

Perihelion Precession

According to Wu's Spacetime Theory, when an object travels close to a massive star, the speed of the object V decreases due to the large Wu's Unit Length l_{yy} caused by the massive gravitational field ($V \propto l_{yy}^{-1/2}$), such that the path of the traveling object bends toward the star in order to maintain its structure coherency. This phenomenon is called "Perihelion Precession".

A schematic explanation can be presented as follows:

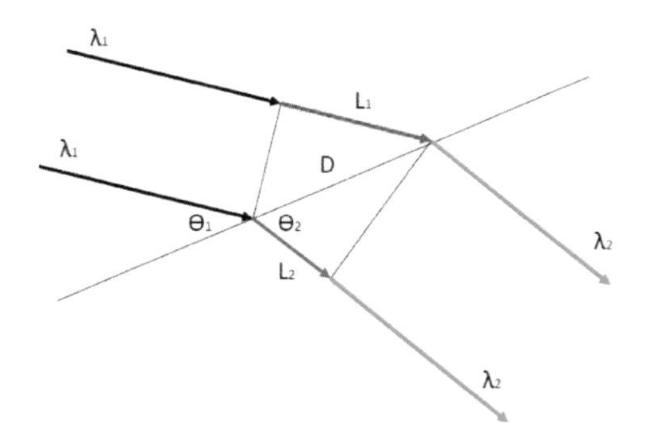

Because

$L_1 = V_1 t = D \cos \Theta_1$

$L_2 = V_2 t = D \cos \Theta_2$

$$V_1 > V_2$$

Therefore,

$L_1 > L_2$

$\cos \Theta_1 > \cos \Theta_2$

$$\Theta_1 < \Theta_2$$

Made in the USA
San Bernardino, CA
10 March 2020

65498784R00144